Environmental Science and Engineering

The ultimate goal of this series is to contribute to the protection of our environment, which calls for both profound research and the ongoing development of solutions and measurements by experts in the field. Accordingly, the series promotes not only a deeper understanding of environmental processes and the evaluation of management strategies, but also design and technology aimed at improving environmental quality. Books focusing on the former are published in the subseries Environmental Science, those focusing on the latter in the subseries Environmental Engineering.

Alexander Geraldovich Arkhipov
Editor

Sustainable Fisheries and Aquaculture: Challenges and Prospects for the Blue Bioeconomy

Springer

Editor
Alexander Geraldovich Arkhipov
Department of Aquatic Bioresources
and Aquaculture
Kaliningrad State Technical University
Kaliningrad, Russia

ISSN 1863-5520 ISSN 1863-5539 (electronic)
Environmental Science and Engineering
ISBN 978-3-031-08283-2 ISBN 978-3-031-08284-9 (eBook)
https://doi.org/10.1007/978-3-031-08284-9

This Springer imprint is published by the registered company Springer Nature Switzerland AG
The registered company address is: Gewerbestrasse 11, 6330 Cham, Switzerland

Introduction

The problems of sustainable fisheries and aquaculture as the basis for the development of the blue bioeconomy are given great attention in international fisheries organizations, as well as in the formation of national development strategies in most countries of the world. A feature of the Blue Bioeconomy Concept is the emphasis on the systematic and sustainable development of economic sectors related to the use of aquatic biological resources. The development of the Concept reflects the recognition of the exceptional role of fresh and salt water for the world community. The rapid degradation of aquatic ecosystems and the depletion of their resource base due to increased environmental and economic pressure inevitably affects the well-being and health of people living in different countries. Despite the growing challenges, the World Ocean still provides mankind with great opportunities for the development of various resources, including the aquatic biological resources. A general consensus on understanding the Blue Bioeconomy and harnessing the potential of the oceans for growth was first reached by countries at the 2012 Rio+20 UN Sustainable Development Conference. In 2015, the idea of the concept formed the basis for the definition of the UN Sustainable Development Goals, in particular, the 14th goal sounds like "conservation and rational use of the oceans, seas and marine resources for sustainable development". The concept of "blue bioeconomy" is aimed at the systematic and sustainable development of economic sectors associated with the use of the resources of the World Ocean, and reflects the desire of the world community to find a balance between further economic growth and development, maintaining environmental safety and ensuring the social well-being of the Earth's population.

The main purpose of this compendium is to promote the implementation of new initiatives and projects in the field of fisheries and aquaculture in the framework of the development of the blue bioeconomy, as well as the exchange of experience in the field of fisheries. One of the ideas of the proposed monograph is to assess new prospects for future cooperation within the framework of projects implemented by the "SUBMARINER Network for Blue Growth platform". The collection contains the following sections: aquaculture, sustainable fishing, ecology and protection of aquatic ecosystems and fisheries. The monograph is intended for specialists in the field of fisheries and aquaculture, ecologists, teachers and students of biological specialties of universities and colleges.

Contents

Part I Aquaculture

1 Modern State and Prospects of Aquaculture Development in the Northwestern Caucasus 3
N. G. Pashinova, A. V. Abramchuk, and G. A. Moskul

2 The Effect of Modern Probiotic Preparations on Enterosorbents on the Piscicultural and Microbiological Parameters of Sturgeon Fish 11
Yu. A. Sergina, A. D. Zhandalgarova, A. A. Bakhareva, Yu. N. Grozesku, Yu. V. Fedorovykh, and A. N. Nevalennyi

3 Pasture Fish Farming in the Reservoirs of the Volga-Kama Cascade as One of the Directions for the Implementation of the Blue Bioeconomy Concept 21
V. A. Shashulovsky, Yu. A. Malinina, and V. V. Kiyashko

Part II Ecology and Protection of Aquatic Ecosystems

4 Research Microplastics—Hydraulic Size of Microplastic Particles of Regular Shape and Their Distribution Over the Depth of the Watercourse 31
N. R. Akhmedova and V. A. Naumov

5 Assessment of the Trophic Status of the Coastal Area of Lake Vistytis (Kaliningrad Region) by Hydrochemical Parameters 41
E. V. Krivopuskova and N. N. Tzvetkova

6 Vistula Lagoon as Potential Object of Accumulated Environmental Harm 53
A. V. Minashkina and S. V. Kondratenko

7 Invasive Species *Palaemon Elegans* Rathke, 1836, (Caridea: Palaemonidae) as the Only Species of Palemon Shrimps in Water Bodies of the Kaliningrad Region 63
S. A. Sudnik and Yu. E. Egorova

8 Influence of Water Quality in Urban Water Reservoirs of Kaliningrad on the State of Amateur Fishing 73
N. A. Tsupikova, E. A. Sevostianova, and A. S. Popova

Part III Fisheries Management

9 Prospects of Development of Small-Scale Fisheries of Vendace in Lake Vistytis, Kaliningrad Oblast, Russia 85
A. V. Aldushin and S. V. Shibaev

10 Rational Use of Aquatic Biological Resources of the Atlantic Ocean for the Sustainable Economic Development of Russia 97
A. G. Arkhipov, V. I. Sauskan, and V. M. Osadchiy

11 Problems of Arrangement on the Private Land Holdings of Ponds, That Have a Hydraulic Connection with Other Water Objects 105
Nina Bugakova, Anna Machanek, and Natalia Yurasyuk

12 Problems of Implementation of the Program for the Development of the Fisheries Industry of the Russian Federation 115
Victoriya V. Dorofeeva, Elena V. Klippenshtein, and Yulia S. Morozova

13 Natural and Artificial Reproduction of Sterlet in the Cheboksary Reservoir 125
A. E. Minin and L. M. Minina

14 Dynamics of Water Levels in the Gorky Reservoir (Russian Federation) During the Spawning Period and Its Influence on the Area of Spawning Grounds and Natural Reproduction of Limnophilic Fish Species 139
L. M. Minina and A. E. Minin

15 Fisheries and Aquaculture: Implementing Sustainable Development Goals 149
Natalia N. Yarkina and Natalia N. Logunova

Part IV Fishing

16 Structure of Catches and Dynamics of Fisheries in the Azov-Black Sea Basin Within Krasnodar Krai 163
A. V. Abramchuk, N. G. Pashinova, and G. A. Moskul

17 Impact of Gidrometerogical Conditions on Smelt Spawning Migration and Catch Fluctuations in the Rivers of Curonian Lagoon Basin ... 171
Anna Burbakh and Sergey Shibaev

18 Characteristics of the Ichthyocenosis in the Wintering Pits in the Context of the Zoogeographic Origin on the Example of Shallow Waters of the Volgograd Reservoir Basin ... 181
Irina A. Kiyashko and Vladimir V. Kiyashko

19 Ways of Penetration and Further Spread of White-Eye Bream (*Ballerus sapa*, Pallas, 1814) in Inland Water Bodies of Kaliningrad Oblast (Russia) ... 191
Oleg Novozhilov, Yulia Aldushina, and Pavel Baranovsky

20 Using the Gill Nets Survey for Assessment of Fish Stock and Allowable Catch in the Vistytis Lake, Kaliningrad Oblast, Russia ... 199
Sergey Shibaev, Oleg Novozhilov, and Pavel Baranovsky

Part I
Aquaculture

Chapter 1
Modern State and Prospects of Aquaculture Development in the Northwestern Caucasus

N. G. Pashinova, A. V. Abramchuk, and G. A. Moskul

Abstract The water resources of the Northwestern Caucasus (Krasnodar krai and the Republic of Adygea) suitable for intensive development of aquaculture are considered. Channel, floodplain and estuary ponds, as well as the Azov-Kuban estuaries, the rivers of the Azov-Kuban plain and reservoirs of the Kuban basin are the most promising for fisheries. They are the resource base for the functioning of different categories of industrial fish-farming enterprises (pond, lake-commodity, basin, farm and cage), which annually produce more than 20 thousand tons of marketable fish products. At the same time, the available water bodies are used inefficiently for fishery purposes. Fish productivity in water bodies ranges from 3 to 1600 kg/ha, averaging 178.6 kg/ha. For the purpose of the rational use of the available natural food resources of the water bodies (the Kuban estuaries, rivers of the Azov-Kuban plain and reservoirs of the Kuban basin), their stocking with yearlings of common carp (Cyprinus carpio), silver carp (Hypophthalmichthys molitrix), bighead carp (Hypophthalmichthys nobilis) and grass carp (Ctenopharyngodon idella) with an individual weight no less than 25–30 g is recommended. Black carp (Mylopharyngodon piceus), so-iuy mullet (Liza haematocheilus), paddlefish (Polyodon spathula) etc. can be used as additional species in aquaculture. Activities aimed at restructuring of commercial ichthyocenoses, together with measures of fishery reclamation, will significantly increase the volume of freshwater aquaculture production.

Keywords Northwestern Caucasus · Rivers · Reservoirs · Estuaries · Ponds · Aquaculture · Fish productivity

1.1 Introduction

The territory of the Northwestern Caucasus, within the Krasnodar krai and the Republic of Adygea, has a significant fund of fishery water bodies suitable for intensive development of aquaculture, the total area of which exceeds 250 thousand ha

N. G. Pashinova (✉) · A. V. Abramchuk · G. A. Moskul
Kuban State University, Stavropol str. 149, Krasnodarad 350040, Russian Federation
e-mail: apilab@yandex.ru

A. G. Arkhipov (ed.), *Sustainable Fisheries and Aquaculture: Challenges and Prospects for the Blue Bioeconomy*, Environmental Science and Engineering, https://doi.org/10.1007/978-3-031-08284-9_1

(Moskul et al. 2018). The natural and climatic factors of the region provide favorable conditions for the development of commercial aquaculture using traditional resource-saving technologies, primarily pond and lake commercial fish farming. Further development of fish-farming enterprises in the inland waters of the region makes it possible to increase the volume of aquaculture production, using the available resources.

1.2 Materials and Methods

The main purpose of this work is to study the current state of the aquaculture sector of the Northwestern Caucasus and to identify promising ways of its development. In accordance with the purpose of the research, the results of the activities of commercial aquaculture enterprises for different categories of farms (pond, lake-commodity, basin, farm and cage) have been analyzed for the period 2019–2020. The inland natural and artificial water bodies of the Northwestern Caucasus such as estuaries, rivers of the Azov-Kuban plain (Eya, Beysug, Chelbas, Kirpili and their tributaries), reservoirs of the Kuban basin (Krasnodar, Kryukov, Varnavinskoe) which are of high fishery importance, were studied (see Fig. 1.1).

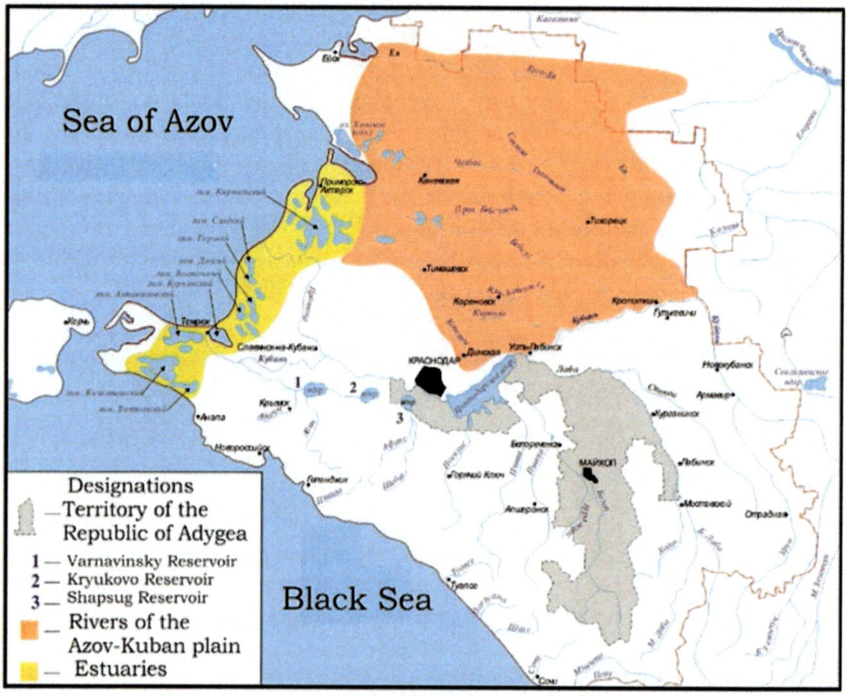

Fig. 1.1 Water resources of the northwestern Caucasus

1.3 Results

The levee-type ponds, as well as the Kuban estuaries, rivers of the Azov-Kuban plain and reservoirs of the Kuban basin are the most promising in regard to commercial fisheries. They are the resource base for the functioning of different categories of commercial aquaculture enterprises (pond, lake-commodity, basin, farm and cage), which annually produce more than 20 thousand tons of marketable aquaculture products. Lake-commodity and pond farms are the most important; their proportion accounts for an average of 66% and 26%, respectively (Fig. 1.2). These fish-farming enterprises specialize in the cultivation of cyprinid fish species: carp (Cyprinus carpio), silver carp (Hypophthalmichthys molitrix), bighead carp (Hypophthalmichthys nobilis) and grass carp (Ctenopharyngodon idella). Annually, these enterprises produce over 13.9 thousand tons of commercial fish. In addition to typical species of freshwater polyculture, farms and basin-type enterprises actively grow sturgeon species (Acipenser ruthenus, Acipenser gueldenstaedtii, Acipenser baerii and their hybrid forms), as well as salmon (Oncorhynchus mykiss, etc.), their share in the volume of commercial aquaculture production is 7.5% and 0.5%, respectively.

Fish farms (FF) are quite widely represented in all water bodies of the region, while, depending on the chosen technology, they organize the fish cultivation according to the pond, cage, lake-commodity or basin principles. The annual volume of fish production is about 900 tons.

The analysis of the available materials shows that draining (levee-type) ponds are the most effectively used within the territory of the Northwestern Caucasus. The fish productivity is on average 1.2 t/ha, and it ranges from 0.4 to 1.6 t/ha for particular ponds. The best results were obtained in fish farms where multiple measures concerning intensification of aquaculture are applied. Along with the multiculture of fish (joint cultivation of carp, herbivorous and other valuable fish species, such as sturgeon, mullets, etc.), feeding of fish and pond fertilization are applied. Polyculture makes it possible to significantly increase the economic efficiency of pond fish

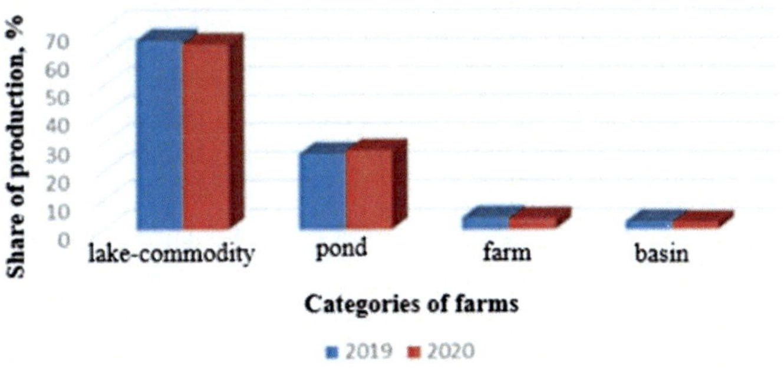

Fig. 1.2 Commercial aquaculture production by different categories of fish farms in the field of industrial fish farming in the northwestern Caucasus (according to the data of the ministry of agriculture and processing industry of Krasnodar krai)

farming by using all food resources of the water body (phytoplankton, zooplankton, zoobenthos, macrophytes and detritus).

The Kuban estuaries, ponds on the rivers of the Azov-Kuban plain, as well as reservoirs of the Kuban River basin, are represented by one group, lake-commodity fish farms (see Fig. 1.2).

The Kuban estuaries have always been reproduction and feeding grounds for valuable fish species in the Sea of Azov (Sander lucioperca, Rutilus heckelii, Vimba vimba, Alburnus chalcoides, etc.). A sharp increase in the sea salinity and the deterioration of the hydrological regime in the estuaries led to a decrease in the reproduction efficiency of the main commercial fish species. Large areas of water bodies with sufficient food resources are used inefficiently. Analysis of the results of the studies shows that when applying lake commodity fish-farming technologies in the estuaries and stocking them with yearlings of common carp (Cyprinus carpio), silver carp (Hypophthalmichthys molitrix), bighead carp (Hypophthalmichthys nobilis) and grass carp (Ctenopharyngodon idella) with an individual weight of 25–30 g, fish productivity will significantly increase with a 50% commercial return and an individual weight of 1.0–1.5 kg (Moskul et al. 2017).

There are a great number of rivers in the Northwestern Caucasus suitable for aquaculture development. More than 10,000 rivers flow through the territory of Krasnodar krai (Borisov 2005). The largest of them are the Eya, Chelbas, Kirpili, and Beysug rivers. The Azov-Kuban rivers and their tributaries are regulated and represent a cascade of small reservoirs and ponds, ranging from several hectares in the upper reaches of the rivers to 300–400 hectares in the lower reaches (Abramchuk et al. 2020a). In total, there are more than 1300 ponds and small reservoirs with a total area of about 49 thousand ha on the rivers of the Azov-Kuban plain. Most water bodies were subject to a significant anthropogenic transformation and not all of them can be used for aquaculture purposes, as they need serious reclamation work.

For the purposes of aquaculture, they can be used after comprehensive reclamation and restoration works:

- clearing of silt and higher aquatic vegetation. This will make it possible to open springs, improve the hydrochemical regime of water and increase the depth of water bodies;
- reconstruction of earthen dams and destruction of small overgrown ponds;
- afforestation of the banks by planting forest belts and limited ploughing of river valleys;
- fishing in reservoirs aimed at removing low-value, slow-growing and predatory fish species.

Implementation of the above-mentioned fish-breeding and reclamation measures and stocking of water bodies with yearlings of common carp (Cyprinus carpio), silver carp (Hypophthalmichthys molitrix), bighead carp (Hypophthalmichthys nobilis) and grass carp (Ctenopharyngodon idella) with an individual weight of 25–30 g will increase the fish productivity to more than 1.1 t/ha in the third year after stocking, taking into account additional and local fish species (Abramchuk et al. 2020).

Several reservoirs were built at different times on the largest rivers of the North-western Caucasus to provide the necessary water supply, primarily for irrigation and rice planting, (Kuban, Belaya, Afips, etc.). In 1941, the Tshchik Reservoir was put into operation which completely regulated the Belaya River and partially regulated the Kuban River flow. In 1952, the construction of the Shapsug reservoir was completed at the mouth of the Afips River. In 1970–1971, the Kryukov and Varnavinsky reservoirs were put into operation. In 1974–1975, the Krasnodar Reservoir, the largest reservoir in the region, was put into operation, the total area of which included the Tshchik Reservoir. The construction of this artificial reservoir completely regulated the Kuban River flow. The total area of reservoirs in the region is 48 thousand ha (Moskul et al. 2018; Abramchuk et al. 2020b).

According to the project data, fisheries in the reservoir should be based on naturally forming stocks of local fish (Cyprinus carpio, Abramis brama, Sander lucioperca). The mode of operation of all reservoirs coincides with the natural reproduction of the main commercial fish species. The studies have demonstrated that the level regime of reservoirs in the Kuban basin is subject to significant fluctuations, especially in the spring period, when the reservoir drawdown can reach an average of 55 cm per decade, which negatively affects the natural fish reproduction. An intensive discharge of water for flooding rice checks occurs during the period of mass spawning of common carp (Cyprinus carpio), bream (Abramis brama) and other phytophilous fish species, which causes the drainage of large spawning areas (up to 70–80%) and the mortality of deposited eggs. In addition, every year the disruption of the oogenesis is observed in 45–60% of female carp and bream, which is accompanied by caviar resorption. All this significantly affects the yield of juveniles and, consequently, the commercial stocks and fish productivity of reservoirs (Abramchuk et al. 2020b).

According to the project-related data, the fish productivity in the reservoirs of the Kuban basin should be about 30 kg/ha, taking into account the annual stocking with juvenile carp and herbivorous fish. According to the forecast data, during the years of intensive stocking with both fingerlings and yearlings and two-year-old fish, it was permitted to catch more than 30 kg/ha in the Krasnodar Reservoir and 45–60 kg/ha in other reservoirs of the Kuban basin. Since the mid-1990s, due to a noticeable decrease in the volume of stocking, catches have not exceeded an average of 2–4 kg/ha in the Krasnodar Reservoir, and 5–6 kg/ha in the other reservoirs (Abramchuk et al. 2020b). The studies have demonstrated that when reservoirs are stocked with yearlings of fish (Hypophthalmichthys molitrix, Hypophthalmichthys nobilis and Ctenopharyngodon idella and additional fish species Cyprinus carpio, Mylopharyngodon piceus, Liza haematocheilus, Polyodon spathula, etc.) with an individual weight of at least 25–30 g, the total fish productivity will be more than 0.2 t/ha in the third year after stocking upon reaching an individual fish weight of 2.0–2.5 kg and a 20% commercial return (Moskul 2017).

1.4 Discussion

As mentioned earlier, the territory of the Northwestern Caucasus, within the Krasnodar krai and the Republic of Adygea has a significant fund of fishery water bodies suitable for aquaculture using different technological approaches. At the same time, it has been established that a significant part of the natural food resources in the water bodies in the region is not fully used (Moskul et al. 2017; Abramchuk et al. 2020b).

For the purpose of the rational use of the available natural forage potential, the Kuban estuaries, ponds on the rivers of the Azov-Kuban plain and reservoirs of the Kuban River basin are recommended for stocking with yearlings of carp (Cyprinus carpio), silver carp (Hypophthalmichthys molitrix), bighead carp (Hypophthalmichthys nobilis) and grass carp (Ctenopharyngodon idella), as well as black carp (Mylopharyngodon piceus), so-iuy mullet (Liza haematocheilus), American paddlefish (Polyodon spathula), etc. Activities aimed at restructuring of commercial ichthyocenoses, together with measures of fishery reclamation, will significantly increase the volume of freshwater aquaculture production.

1.5 Conclusion

The results of the studies conducted in the water bodies of the Northwestern Caucasus show that forage organisms are not fully used. For the most complete use of residual feed, it is necessary to stock these reservoirs with yearlings of carp (Cyprinus carpio), silver carp (Hypophthalmichthys molitrix), bighead carp (Hypophthalmichthys nobilis), and grass carp (Ctenopharyngodon idella). The rational use of the natural food resources will significantly increase the volume of freshwater aquaculture production.

References

Abramchuk A, Moskul G, Pashinova N, Shumeyko D, Tyurin V (2020a) Fish farming on riverbed ponds of the Azov-Kuban plain. In: XIII International Scientific and Practical Conference "State and Prospects for the Development of Agribusiness—INTERAGROMASH 2020a". INTERAGROMASH, Rostov-on-Don 02011 DOI:https://doi.org/10.1051/e3sconf/202017502011

Abramchuk A, Moskul G, Pashinova N, Yurina Na, Yurin D (2020b). Pastoral aquaculture in the reservoirs of the Kuban basin. In: XIII International Scientific and Practical Conference "State and Prospects for the Development of Agribusiness—INTERAGROMASH 2020b". Rostov-on-Don, Russian Federation. 02010 https://doi.org/10.1051/e3sconf/202017502010

Borisov VI (2005) Rivers of Kuban. Kuban Publishing House, Krasnodar (in Russian)

Moskul GA, Abramchuk AV, Pashinova NG (2017) The current state and prospects for the development of pasture fish farming in the Krasnodar krai. In: All-Russian Scientific Conference.

Aquaculture: world experience and Russian developments. 2017. Southern Scientific Center of the Russian Academy of Sciences, Rostov-on-Don, pp. 474–477 (in Russian)
Moskul GA, Abramchuk AV, Pashinova NG (2018) Prospects of fishery development in inland water bodies of the Krasnodar krai. In: All-Russian Scientific and Practical Conference. Aquatic Bioresources and Aquaculture of the South of Russia. KubGU, Krasnodar, pp. 180–184 (in Russian)

Chapter 2
The Effect of Modern Probiotic Preparations on Enterosorbents on the Piscicultural and Microbiological Parameters of Sturgeon Fish

Yu. A. Sergina, A. D. Zhandalgarova, A. A. Bakhareva, Yu. N. Grozesku, Yu. V. Fedorovykh, and A. N. Nevalennyi

Abstract Currently, in order to prevent and treat a large number of pathological conditions caused by a violation of endoecology, enterosorbent-based preparations that help cleanse the body of metabolic products, heavy metals, radionuclides and various toxins have been widely used. However, the presence of only enterosorbents in the preparation can have a negative impact on the body, since these substances eliminate not only toxins, but also metabolites necessary for the full functioning of the body, and also reduce the number of beneficial microorganisms. To solve this problem, it is necessary to use complex preparations, in particular probiotics on enterosorbents, which have increased efficiency of sorption and desorption of eubiotic bacteria. Microorganisms located in the pores, as well as microbial cells located close to the outer surface of the sorbent, differ in their desorption properties. The porous carrier, being an enterosorbent, relieves the effects of toxicosis, which contributes to the survival of the bacteria of the drug and the intestinal microflora, and also reduces the load on the detoxification organs of fish. The presence of bifidobacteria and lactobacilli in the preparation contributes to an increase in the rate of biomass accumulation in the conditions of the technological process by preserving enzymatic activity and increasing the efficiency of the use of nutrient media. It was found that the introduction of a probiotic drug immobilized on an enterosorbent,

Yu. A. Sergina (✉) · A. D. Zhandalgarova · A. A. Bakhareva · Yu. N. Grozesku · Yu. V. Fedorovykh · A. N. Nevalennyi
Astrakhan State Technical University, 16, St. Tatishcheva, 414056 Astrakhan, Russia
e-mail: 0954632@bk.ru

A. A. Bakhareva
e-mail: bahareva.anya@yandex.ru

Yu. N. Grozesku
e-mail: grozesku@yandex.ru

Yu. V. Fedorovykh
e-mail: jaqua@yandex.ru

A. N. Nevalennyi
e-mail: rector@astu.org

A. G. Arkhipov (ed.), *Sustainable Fisheries and Aquaculture: Challenges and Prospects for the Blue Bioeconomy*, Environmental Science and Engineering, https://doi.org/10.1007/978-3-031-08284-9_2

"Ecoflor" in a dosage of 4 g/kg of compound feed contributes to an increase in fish-breeding biological and microbiological indicators of farmed fish. There will be an improvement in the physiological state of fish and an increase in the protective properties of the immune system. "Ecoflor" has pronounced sorption properties, the ability to fix and remove bacteria, toxins and allergens from the intestine.

Keywords Russian sturgeon · Probiotic preparation · Enterosorbent · Fish-breeding and biological indicators · Microbiological indicators

2.1 Introduction

Currently, it is urgent to search for new approaches to the maintenance and cultivation of sturgeon species of fish, allowing to obtain the maximum yield of products, spending the minimum amount of resources. One of such directions is the normalization of the bacterial flora of hydrobionts grown in closed water supply installations, based on the use of various kinds of probiotic preparations.

It is proved that the density of the microbial population of the contents of the gastrointestinal tract of sturgeon fish depends on the density of the microbial population in the reservoir. Even during periods of wintering and starvation, the intestines of fish are not freed from bacterial population, however, the number of bacteria decreases by 10 times or more in comparison with their number with intensive nutrition.

The survival and viability of all fish species directly depends on their diet. The cultivation of sturgeon species of fish is based on a properly formulated feeding technology, and this is what ensures the maximum survival rates and growth rate, while spending the minimum amount of feed.

To date, it is necessary to create recipes for highly effective compound feeds for fish, including sturgeon species, and for this you need to carefully study the nutrients of natural fish food and the characteristics of all body systems. Also, to create a complete feed, it is necessary to select all the components correctly, while studying not only their production properties, but also to fully balance the feed in terms of nutrient content, based on the needs of the cultivated species at a certain stage of development.

2.2 Materials and Methods

A promising direction in feed production is the production of compound feeds with the use of probiotic preparations on enterosorbents necessary for the prevention and treatment of bacterial and viral etiology of fish diseases, normalization of intestinal microflora by removing pathogenic and conditionally pathogenic microorganisms, toxins and metabolic products from the body (Zhandalgarova et al. 2017). Such preparations can mitigate the stresses caused by the change of compound feeds and technological effects on the body of fish, serve to increase the resistance of the body and the tension of immunity, allow to increase the digestibility of compound feeds (Zhandalgarova 2014).

This trend in the development of feed production makes it necessary to conduct scientific research, the purpose of which is to evaluate the effectiveness of the use of probiotic preparation on enterosorbent in production compound feeds for sturgeon fish.

To achieve the goal, the following tasks were set:

(1) To study the piscicultural indicators of two-year-old Russian sturgeon when grown on mixed feed with the addition of enterosorbent;
(2) Analyze the microflora of the gills and intestines of the Russian sturgeon when feeding with compound feed with the addition of the "Ecoflor" preparation.

Experimental studies were conducted at the Innovation Center "Bioaquapark—Scientific and Technical Center of Aquaculture" of the Federal State Educational Institution of Higher Education "ASTU". The objects of research were two-year-old Russian sturgeon (Acipenser gueldenstaedtii Brandt et Ratzeburg, 1833).

For the research, the probiotic preparation "Ecoflor" was used, which is a consortium of natural strains of lacto- and bifidobacteria: B. bifidum, B. longum and L. acidophilus, L. casei, L. plantarum, immobilized on the carbon-mineral sorbent SUMC-1. The normative content of bifidobacteria in the preparation is 108 CFU / g, lactobacilli—107 CFU / g.

Throughout the experimental work, fish were kept in fiberglass tanks with rounded corners with a volume of 1 m^3. At the same time, the stocking density was 15 individuals per tank. The content of dissolved oxygen in water during the entire growing period ranged from 7–9 mg/ l, pH—6.5–7.0, and the water temperature was maintained at 18–22° C.

Studies on the effectiveness of the use of a probiotic preparation as part of production compound feeds were carried out in laboratory conditions for 30 days. To do this, the fish were divided into 4 groups. The control group was fed with compound feed of the OT-7 formulation. Feeding of the first experimental group was carried out with compound feed OT-7 with the addition of the "Ecoflor" preparation at a dosage of 2 g per 1 kg of feed, the second experimental group—with the addition of the preparation at a dosage of 4 g per 1 kg of feed, the third experimental group—with the addition of 6 g of the preparation per 1 kg of feed.

Throughout the experiment, the daily feeding rate was calculated based on data on the body weight of fish and the water temperature in the tanks, in accordance with the generally accepted cultivation technology (Ponomarev et al. 2002).

The size and weight parameters of the studied individuals of the Russian sturgeon were determined by the method of Pravdin I.F. (1966).

In order to assess the symbiotic microflora of gills and intestines of farmed fish, their microbiological analysis was carried out. Before the selection of fragments of gills and intestines, the fish body was treated with alcohol. The autopsy was performed with sterile instruments. The test samples were placed in pre-prepared sterile penicillin vials. The microflora was studied according to the method of I.V. Burlachenko and L.I. Bychkova (2005). Sampling was carried out three times: 1 sample—before feeding with experimental compound feed with the addition of the "Ecoflor" preparation, 2 sample—in the middle of the experiment (during feeding with experimental compound feed), 3 sample—after feeding. After sampling of the fish under study, seeding was carried out on various media: cabbage agar, Bleakfiltd, Li, nutrient agar (PA) using the method of deep seeding. To analyze the microbiological seeding of the gills and intestines of fish, the method of deep seeding on various media was used. The crops were incubated for 5 days at a temperature of 30° C. The number of microorganisms was determined based on 1 g.

2.3 Results

The effectiveness of feed with the addition of the "Ecoflor" in the diet of two-year-old Russian sturgeon was evaluated on the basis of its feeding for 30 days. The grown individuals were divided into 4 groups: 3 experimental and 1 control. The control group was fed OT-7 compound feed throughout the experiment, 1 experimental group was fed OT-7 compound feed with the addition of the probiotic "Ecoflor" with a dosage of 2 g/kg of feed, 2 experimental group—4 g/kg, 3 experimental group—6 g/kg (Table 2.1).

For the entire experimental period of cultivation, the best results in linear-weight indicators are characteristic of the 2 experimental group, in which the dosage of the drug was 4 g/kg. The absolute increase in the second experimental group was 70,1 g, which is 34 g more than the increase in the control variant.

It should be noted that feeding two-year-old Russian sturgeon with OT-7 feed in the dosage of the "Ecoflor" 4 g/kg allowed to reduce feed costs to 1,2 units, as well as increase the growth rate.

Table 2.1 Piscicultural indicators of growing two-year-old Russian sturgeon

Indicator	Group			
	Control	Test 1	Test 2	Test 3
Initial weight, g	575,3 ± 5,6	564,2 ± 8,2	567,1 ± 8,7	570,5 ± 5,5
Final weight, g	611,4 ± 0,3	632,4 ± 6,9	637,8 ± 9,2	629,5 ± 2,5
Absolute growth, g	36,1	68,2	70,1	59
Average daily growth, g	1,20	2,27	2,34	1,97
Average daily growth rate, %	0,20	0,38	0,39	0,33
Mass accumulation coefficient, units	0,017	0,032	0,033	0,038
Feed ratio	1,5	1,3	1,2	1,3
Survival rate, %	100	100	100	100
Duration of the experiment, day	30	30	30	30

Table 2.2 Data on the quantitative analysis of microorganisms on the gills of experimental individuals of Russian sturgeon before feeding with compound feed with the addition of the "Ecoflor"

Name of the medium	Control group, CFU/g	Test group, CFU/g
Cabbage agar	$2{,}81*10^4$	$2{,}75*10^5$
Bleakfeltd	$4{,}36*10^5$	$3{,}98*10^5$
Li	$3{,}54*10^4$	$2{,}62*10^5$
PA	$1{,}1*10^5$	$1{,}5*10^5$

Throughout the experiment, the piscicultural indicators of two-year-old Russian sturgeon raised in the second experimental group were significantly better than in the control variant. At the same time, if we compare the results in the three experimental groups, then it was in the second that there was a rapid improvement in linear-weight characteristics.

During the experimental cultivation, a qualitative and quantitative analysis of bacterial microflora on the gills of the Russian sturgeon was carried out before and after feeding with a probiotic preparation immobilized on the enterosorbent "Ecoflor". The method of deep seeding on various media was used: cabbage agar, Bleakfeltd, Li and PA (Table 2.2). The individuals of Russian sturgeon were divided into two groups: control and experimental. Feeding of the experimental group was carried out with compound feed with the addition of 4 g/kg of the probiotic drug "Ecoflor", since this dosage is the most optimal.

During the study of isolated cultures of microorganisms before the start of experimental studies, the following colonies were identified: colonies of beige-milky color, slimy, rounded; colonies of yellow color, rounded, glossy; colonies of milky color, creeping along the substrate; colonies of brown color, irregular shape; colonies of pale pink color, round, convex glossy. Among the above colonies, gram-positive bacillus (group 19; Lactobacillus), gram-positive cocci (group 17; p. *Pedioccocus, Leuconoctoc*) and gram-positive spore-forming bacillus (group 18; p. *Bacillus*) (Bergey 1997).

After the end of feeding the experimental group with compound feed with the addition of a probiotic preparation immobilized on enterosorbent, an additional analysis of the quantitative and qualitative microbiological state of the gills of experimental individuals of Russian sturgeon was carried out. Samples were also taken from the control and experimental groups. The method of deep seeding on various media was used (Table 2.3).

During the study of the isolated cultures of microorganisms after the completion of feeding with compound feed with the addition of the "Ecoflor", the following colonies were identified: colonies that eat away the substrate; colonies of orange color, rounded, wrinkled, glossy; colonies of milk color, irregular shape, matte; colonies of beige-milk color, slimy, rounded; colonies of brown color, oval, convex; colonies of pale pink color, round, convex glossy. As a result of the analysis of

Table 2.3 Data on the quantitative analysis of microorganisms on the gills of experimental individuals of Russian sturgeon after the end of feeding with the addition of the "Ecoflor"

Name of the medium	Control group, CFU/g	Test group, CFU/g
Cabbage agar	$1{,}57*10^4$	$2{,}75*10^5$
Bleakfeltd	$1{,}57*10^5$	$1{,}63*10^5$
Li	$1{,}42*10^4$	$1{,}62*10^5$
PA	$1{,}42*10^6$	$1{,}5*10^6$

microbiological seeding, lactic acid microorganisms were identified: gram-positive bacillus and cocci belonging to the genera *Lactobacillus, Pedioccocus, Leuconoctoc.*

When feeding two-year-olds of Russian sturgeon with compound feed with the addition of the "Ecoflor" at a dosage of 4 g/kg, an increase in the concentration of microorganisms on the gills was observed.

At the second stage of experimental work, a qualitative and quantitative analysis of the bacterial microflora in the intestines of the Russian sturgeon was carried out before and after feeding with a probiotic immobilized on the enterosorbent "Ecoflor". The method of deep seeding on various media was used: cabbage agar, Bleakfeltd, Li and PA (Table 2.4).

When analyzing the microbiological seeding of the intestines of experimental individuals of Russian sturgeon, several varieties of bacterial colonies were found: milky colonies, creeping along the substrate; beige-milky colonies, slimy, rounded; yellow colonies, rounded, glossy; brown colonies, irregular shape; pale pink colonies, round, convex glossy. Among them, gram-positive bacillus (group 19; Lactobacillus); gram-positive cocci (group 17; p. *Pedioccocus, Leuconoctoc*) and gram-positive spore-forming bacillus (group 18; p. *Bacillus*) (Bergi 1997).

After the experimental group was fed with compound feed with the addition of the probiotic "Ecoflor" immobilized on enterosorbent, an additional analysis of the quantitative and qualitative microbiological state of the intestines of experimental individuals of Russian sturgeon was carried out. Samples were also taken from the control and experimental groups. The method of deep seeding on various media was used (Table 2.5).

During the study of the cultural properties of microorganisms identified on crops, the following bacterial colonies were found: colonies of a corroding substrate; colonies of milk color, irregular shape, matte; colonies of beige-milk color, slimy,

Table 2.4 Data on the quantitative analysis of microorganisms in the intestines of experimental individuals of Russian sturgeon before feeding with compound feed with the addition of the "Ecoflor"

Name of the medium	Control group, CFU/g	Test group, CFU/g
Cabbage agar	$2{,}55*10^4$	$2{,}9*10^5$
Bleakfeltd	$3{,}02*10^5$	$2{,}85*10^5$
Li	$3{,}12*10^4$	$4{,}63*10^5$
PA	$2{,}38*10^5$	$3{,}2*10^7$

Table 2.5 Data on the quantitative analysis of microorganisms in the intestines of experimental individuals of Russian sturgeon after the end of feeding with compound feed with the addition of the "Ecoflor"

Name of the medium	Control group, CFU/g	Test group, CFU/g
Cabbage agar	$3{,}9*10^5$	$1{,}38*10^6$
Bleakfeltd	$1{,}85*10^5$	$3{,}45*10^5$
Li	$1{,}63*10^6$	$3{,}4*10^6$
PA	$1{,}2*10^7$	$5{,}8*10^7$

rounded; colonies of orange color, rounded, glossy; colonies of brown color, oval, convex; colonies of orange color, wrinkled; colonies of milk color, irregular shape, matte; colonies of brown color, oval, convex, rough; pale pink colonies, round, convex glossy; colonies are pale pink, round, convex glossy. In the course of the study of the data obtained as a result of microbiological analysis of the intestines of experimental individuals of Russian sturgeon, lactic acid microorganisms were identified: gram-positive bacillus and cocci belonging to the genera: *Lactobacillus, Pedioccocus, Leuconoctoc* (Bergi 1997).

It should be noted that feeding two-year-olds of Russian sturgeon with compound feed with the addition of the "Ecoflor" had a beneficial effect on the intestinal biocenosis. An increase in the quantitative composition of the bacterial biota of the intestine was observed both during feeding with experimental compound feed and after feeding.

2.4 Discussion

When an immobilized probiotic preparation is introduced into the production feed at a dosage of 4 g/kg, an improvement in piscicultural indicators is observed in the reared individuals of Russian sturgeon (the absolute increase was 70,1 g, the average daily growth rate was 0,39%, the mass accumulation coefficient was 0,33 units), a reduction in feed costs to 1,2 units.

As a result of the analysis of microbiological seeding of gills, lactic acid microorganisms were identified: gram-positive bacillus and cocci belonging to the genera *Lactobacillus, Pedioccocus, Leuconoctoc*. When feeding two-year-olds of Russian sturgeon with compound feed with the addition of the "Ecoflor" at a dosage of 4 g/kg, an increase in the concentration of microorganisms on the gills was observed when feeding with experimental compound feed, immediately after the end of feeding, a decrease in this indicator was observed.

During the study of the data obtained as a result of microbiological analysis of the intestines of experimental individuals of Russian sturgeon, lactic acid microorganisms were identified: gram-positive bacilluus and cocci belonging to the genera:

Lactobacillus, Pedioccocus, Leuconoctoc. It should be noted that feeding two-year-olds of Russian sturgeon with compound feed with the addition of the "Ecoflor" had a beneficial effect on the intestinal biocenosis. An increase in the quantitative composition of the bacterial biota of the intestine was observed both during feeding with experimental compound feed and after feeding.

If an infectious pathology occurs as a result of a violation of the balance between normal and pathogenic intestinal microflora, antibacterial therapy is used. This method involves the use of various kinds of antibiotics. If antibacterial therapy does not bring proper results, but on the contrary increases the load on the animal's body, it is necessary to immediately introduce probiotics into the diet, but the use of probiotics as the main medicine is excluded.

The addition of the drug "Ecoflor" allows you to influence a wide range of problems that are directly related to the physiological state of animals. It is also possible to stimulate the growth and development of the body, while increasing the efficiency of digestion. From which it follows that this preparation is an effective element in the production technology of fish farming products that are safe for humans.

2.5 Conclusion

As a result of the conducted research, it was found that it is advisable to introduce a probiotic on the enterosorbent "Ecoflor" into the production compound feeds for sturgeon fish. The rate of probiotic input is 4 g/kg of compound feed. This dosage contributes to an increase in piscicultural and hematological parameters of fish, as well as to the improvement of the microbiological status of farmed fish.

The use of Russian sturgeon with productive feeds in combination with immobilized probiotic preparations can significantly increase profits and minimize feed costs.

Production compound feed with the addition of an immobilized form of probiotic preparation for sturgeon fish species helps to eliminate the existing negatives in fish farming, as it allows to increase the volume of products obtained, improve the quality of compound feeds. Therefore, it is one of the ways to increase the efficiency of cultivation and rational use of compound feeds.

References

Burlachenko KV, Bychkova LK (2005) Method of clinical assessment of the condition of sturgeon fish during their cultivation in installations with a closed water supply cycle. Fisheries 6:70–72 (in Russian)

Hoult J, Krig N, Snitt P (eds) (1997) The determinant of bacteria Bergi. m The determinant of bacteria Berg. Mir., Moscow, Russia (in Russian)

Ponomarev SV, Gamygin EA, Nikonorov SI, Ponomareva EN, Grozesku YuN, Bakhareva AA (2002) Technologies of cultivation and feeding of aquaculture objects in the South of Russia. Nova Plus, Astrakhan, Russia (in Russian)

Pravdin PF (1966) Guide to the study of fish. Food industry, Moscow, Russia (in Russian)

Zhandalgarova AD (2014) The use of probiotic drugs as sorbents in the toxic effect of surfactants on sturgeon. In: Matishov GG (eds) Proceedings from: topical issues of fisheries and aquaculture of the basins of the southern seas of Russia. Publishing House of the SSC RAS, Rostov-On-Don, Russia (in Russian)

Zhandalgarova AD, Bakhareva AA, Grozesku YuN, Pravdin AI (2017) The influence of a new generation synbiotic feed additive "PROSTOR" on the fish-breeding, biological and hematological parameters of the young Russian sturgeon. Chief Zootechnik 7:30–35 (in Russian)

Chapter 3
Pasture Fish Farming in the Reservoirs of the Volga-Kama Cascade as One of the Directions for the Implementation of the Blue Bioeconomy Concept

V. A. Shashulovsky, Yu. A. Malinina, and V. V. Kiyashko

Abstract In the second half of the twentieth century, many rivers were subjected to the most powerful of all known anthropogenic impacts—channel regulation. This led to the creation of a cascade of large reservoirs on the Volga River, which radically changed the hydrological, hydrochemical and hydrobiological regimes of water bodies and became the most significant for the ecosystems of the basin and natural landscape complexes. At the same time, as practice has shown, the multifunctionality of the reservoirs, unfortunately, does not allow to fully realize their potential fish productivity. Currently, there are forage resources in the reservoirs of the Volga-Kama cascade, which are either not used at all, or are poorly used by representatives of the existing ichthyofauna. The main task of the research was to assess the value of the commercial fish stock and the production of the links of the food chain, which is transformed by fish, and to determine at the expense of which links it is possible to increase the volume of fish production and its withdrawal without violating the functional integrity of the ecosystem. The dynamics of the aggregate commercial stock of the main fish species of the Volga-Kama cascade of reservoirs in recent years does not have a clearly pronounced trend, which may indicate the relative production stability of the ichthyocenosis. The calculation of the receiving capacity of the reservoirs of the cascade shows the possibility of increasing the catch by 18 thousand tons, while the cost of the expected fish products will more than 4 times exceed the cost of the planting material. The creation of a complex of pasture fish farms will have a beneficial effect on the ecological state of water bodies, will make it possible to use the fodder base more fully and efficiently, will ensure biological reclamation, and will make it possible to obtain additional, high-quality fish products.

Keywords Volga river · Commercial fish stock ABR · Catch · TAC · Formation of fish fauna · Receiving capacity · Pasture fish farming

V. A. Shashulovsky (✉) · Yu. A. Malinina · V. V. Kiyashko
Saratov Branch of the "Russian Federal Research Institute of Fisheries and Oceanography", Chernyshevskogo str. 152, 410002 Saratov, Russian Federation
e-mail: saratovniro@vniro.ru

A. G. Arkhipov (ed.), *Sustainable Fisheries and Aquaculture: Challenges and Prospects for the Blue Bioeconomy*, Environmental Science and Engineering, https://doi.org/10.1007/978-3-031-08284-9_3

3.1 Introduction

One of the main tasks put forward by the Blue Bioeconomy Concept is the rational use of natural resources of inland water bodies, an increase in productivity, and obtaining the best results of fishing, both in qualitative and quantitative terms.

The fishery fund of freshwater reservoirs of the Russian Federation includes lakes with a total area of 22.5 million hectares, and reservoirs with an area of 8.9 million hectares and rivers with a length of 523.4 thousand km. The European part accounts for 5.6 million hectares of lakes, 6.0 million hectares of reservoirs (Sechin 2010).

Reservoirs are mainly reservoirs for complex purposes. They are transport arteries, sources of electricity, are used for water supply, irrigation, recreation and other purposes. In addition, the newly created reservoirs were a new fishery fund, the rational use of which, according to design estimates, should have brought at least 300 thousand tons of marketable fish products per year. At the same time, as practice has shown, the multifunctionality of the reservoirs, unfortunately, does not allow to fully realize their potential fish productivity. The formation of fish resources in these reservoirs is largely determined by their artificial origin and the functioning of ecosystems under conditions of unrelenting anthropogenic pressure (Mosiyash and Shashulovsky 2019).

The reservoirs of the Volga-Kama cascade are the most important freshwater fishery reservoirs in Russia. These are 12 reservoirs with a total area of 2.66 million hectares (Kudersky 1984). They are located in a densely populated region of the country with an intensively developed industry and a high concentration of urban population.

Thus, the purpose of this work was to study the possibility of increasing the volume of extraction of valuable fish products in the reservoirs of the Volga-Kama cascade based on a comprehensive analysis of the state of ecosystem components.

3.2 Materials and Methods

The material for the research was the results of a comprehensive study of the components of the ecosystem of the Volga-Kama cascade of reservoirs in 2007–2020.

The informational basis is based on long-term works of authors, literature data and fund materials of the Saratov branch of «VNIRO» and branches of the Volga-Caspian cascade (TatarstanNIRO, PermNIRO, NizhegorodNIRO, «VNIIPRH»).

The selection and processing of hydrobiological material was carried out according to generally accepted methods (Morduhai-Boltovskoi 1975). The feeding capacity of the reservoir was assessed according to the classification of M.L. Pidgayko et al. (1968). Statistical data processing was carried out by conventional methods.

Collection and processing of ichthyological material was carried out according to generally accepted methods (Sechin 1990; Yudovich 1982; Rudenko 1985).

The volume of output was calculated using the equation by Rudenko (1983).

The taxonomic position of fish is given by the identification guides (Reshetnikov 1998, 2002).

3.3 Results

The dynamics of the aggregate commercial stock of the main fish species of the Volga-Kama cascade of reservoirs in recent years does not have a clearly pronounced trend (see Fig. 3.1), which may indicate the relative production stability of the ichthyocenosis (Shashulovsky et al. 2018).

Rational fisheries management primarily relies on an assessment of the size of commercial fish stocks (Shahulovsky et al. 2018) and on the correspondence between this value and the intensity of fishing (see Fig. 3.2).

The transformation of the Volga and Kama into a cascade of reservoirs created favorable conditions for increasing the fish productivity of these large rivers. However, the annual indicators of fish catches are increasing due to economic and organizational measures carried out recently, and not due to an increase in natural fish productivity.

The data obtained indicate that the stocks of native fish species are used as fully as possible.

The integrated use of water resources is associated with a change in the prevailing conditions for the reproduction and habitation of fish and with a violation of the

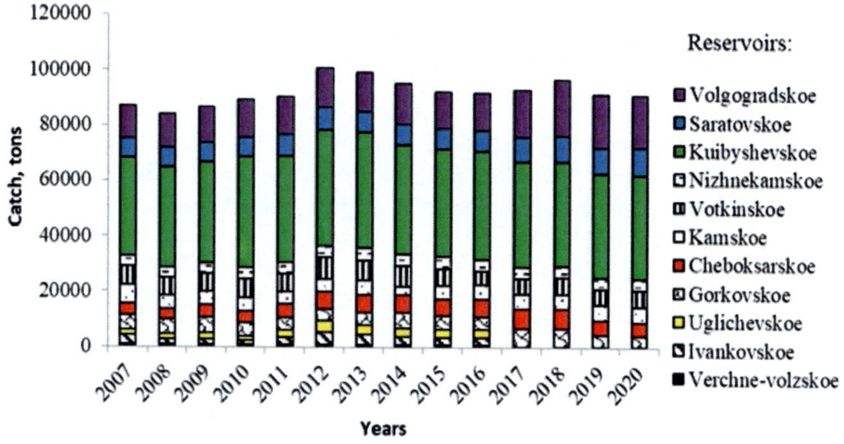

Fig. 3.1 Dynamics of industrial reserves of ABR in the reservoirs of the Volga-Kama cascade. *Source* compiled by the authors

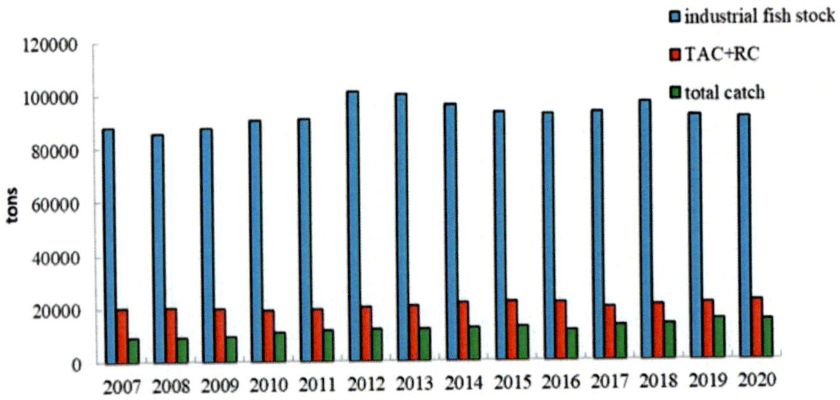

Fig. 3.2 Characteristics of the stock, TAC and the ABR fishery in the reservoirs of the Volga-Kama cascade. *Source* compiled by the authors

complex mechanism of reproduction of fish stocks (Shahulovsky and Mosiyash 2005). The main trends in the formation of the fish population in reservoirs:

- Over the past decades, the share of large predators (catfish and pike perch) has decreased by 3–10 times;
- The ichthyomass of small-sized low-productive species (perch, silver carp, roach, etc.) has increased by 1.5–2 times;
- A significant share of food products of aquatic organisms is consumed by low-value commercial and "weedy" non-commercial species (tulka, bleak, ruff, gobies, etc.);
- Overgrowth of shallow water areas and increased eutrophication.

Currently, the development of fisheries provides for a rational economy and active human influence on the processes of fish reproduction.

Prerequisites for the targeted formation of ichthyofauna:

- The presence of significant forage resources in reservoirs indicates reserves for an increase in commercial fish stocks;
- This can be done by introducing valuable highly productive species: consumers of higher aquatic vegetation, phytoplankton, detritus, mollusks (white and variegated silver carp, white and black carp, etc.);
- The introduction of these fish species becomes especially important in the modern conditions of the fishery "degradation" of the ichthyofauna of large water bodies.

Our studies have confirmed the presence of significant unused forage resources, higher aquatic vegetation, phytoplankton, molluscs, which indicates the urgency of the task of increasing commercial fish stocks by introducing valuable and highly productive fish species: sterlet, grass carp, silver carp, smallmouth buffalo, black carp, carp, carp for the purpose of carrying out pasture fish farming in two directions:

(1) *Foraging*—based on fish species that are not capable of natural reproduction in conditions of stocked reservoirs (herbivorous fish of the Far East complex).
(2) *Reproductive*—based on valuable native species that have lost their reproductive potential due to a number of anthropogenic factors (sturgeon, catfish, pike perch, carp, etc.).

The calculation of the receiving capacity of the reservoirs of the cascade shows the possibility of transforming the energy of trophic "dead ends" into high-quality fish products and increasing the catch (Table 3.1).

The expected result of the introduction of valuable fish species will amount to almost 18 thousand tons of products (see Table). Analysis of the cost of planting material, taking into account the volume of introduction and the commercial value of the harvested fish products, showed more than 100% profitability.

Table 3.1 Results of the calculated receiving capacity of a number of reservoirs in the cascade

Reservoirs of the Volga cascade	The volume of introduction, th. spec	Expected Result, th.t
Ivankovskoe reservoir (sterlet, carp) *	444	0,13
Gorkovskoe (sterlet, carp, catfish, silver carp (white and variegated), pike perch, pike) *	25,750	3,60
Cheboksarskoe (sterlet, carp, catfish, silver carp (white and variegated), pike perch, pike) *	23,850	3,90
Kuibyshevskoe (sterlet, carp, silver carp (white and variegated), grass carp) *	2043	0,30
Nizhnekamskoe reservoir (sterlet, carp, silver carp) *	1500	0,54
Volgogradskoe (sterlet, carp, catfish, silver carp (white and variegated), cupids (white and black)	87,500	6,00
Saratovskoe reservoir (sterlet, carp, catfish, silver carp (white and variegated), grass carp	24,651	3,48
TOTAL:	165,738	17,95

*—materials provided TatarstanNIRO, PermNIRO, NizhegorodNIRO, «VNIIPRH»

3.4 Discussion

The creation of a complex of pasture fish farms will have a beneficial effect on the ecological state of water bodies, will make it possible to use the fodder base more fully and efficiently, will ensure biological reclamation, and will make it possible to obtain additional, high-quality fish products.

The main reasons for the slow development of pasture fish farming and artificial reproduction activities:

- Lack of viable fish seed;
- Lack of fish hatcheries and breeding complexes or their insufficient capacity;
- Underestimation of scientific support and support of events;
- Lack of qualified personnel;
- Lack of a system for collecting reliable information on pasture aquaculture products.

Increasing the efficiency of pasture fish farming and artificial reproduction of aquatic biological resources should be accompanied by:

- carrying out a comprehensive cadastral study of reservoirs in federal ownership;
- improving organizational procedures and reproduction processes;
- development and implementation of resource-saving technologies;
- reconstruction and technical re-equipment of existing enterprises for artificial reproduction;
- construction of new reproduction complexes in order to overcome the shortage of planting material.

3.5 Conclusion

An assessment of the possibilities and determination of the development potential of the blue bioeconomy of reservoirs formed on the largest European river showed that the natural biological resources of the Volga-Kama reservoir basin of the cascade at the present time, when conducting an extensive economy, do not significantly increase the production of valuable fish products.

The main direction of the development of the biological resource base in the basin of reservoirs should be aquaculture, focused on increasing the number and expanding the range of fish products.

In the future, the basis for the fishery exploitation of the Volga reservoirs should be their transfer to the mode of pasture aquaculture with the simultaneous development of cultural recreational fishing.

To maintain and preserve the fishery status of reservoirs, it is necessary to create and operate specialized subdivisions for their fishery reclamation.

Thus, in order to transform the energy of trophic "dead ends" into high-quality fish products, increase the catch, reduce the process of eutrophication of the reservoir, as well as restore the number of valuable fish species (sterlet, catfish, and carp), artificial reproduction and annual release of viable planting material in scientific –justified volumes. This will allow preserving the biodiversity of aboriginal fish species, significantly reduce the load on the aquatic fish fauna, create additional jobs in the fishing and fish processing industries, and reduce social tension in the regions.

References

Kudersky LA (1984) Status and Ways of Development of Fisheries in the Reservoirs of the Volga-Kama Cascade Sbornik Nauchnykh Trudov GosNIORKH 217:3–17 (In Russian)
Morduhai-Boltovskoi FD (Ed.). (1975). Methods of studying biogeocenoses of inland water bodies. Moscow, Russia: Nauka (In Russian)
Mosiyash SS, Shahulovsky VA (2019) Patterns of the dynamics of commercial fish stocks in the macroecosystem of the Volga-Kama cascade. KMK Scientific Press, Moscow, Russia (In Russian)
Pidgaiko ML, Aleksandrov BM, TsI I (1968) Brief biological and production characteristics of water bodies in the North-West of the USSR. Izvestiya GosNIORKH 67:205–228 (In Russian)
Reshetnikov YuS (Ed.). (1998). Annotated Check-List of Cyclostomata and Fishes of the continental Waters of Russia. Moscow, Russia: Nauka (In Russian)
Reshetnikov YuS (Ed.). (2002). Atlas of Russian Freshwater Fishes (2nd ed.) (vol. 1). Moscow, Russia: Nauka, p. 378 (In Russian)
Rudenko GP (1985). Methods for determining the ichthyomass, the growth of fish and fish products. In G.G. Vinberg & M.I. Yarushina (Eds.), Production of populations of aquatic organisms communities and methods for its study (p. 111–138). Sverdlovsk: UNTS AN SSSR (In Russian)
Rudenko GP (Ed.). (1983). Handbook of lake and cage fish farming. Moscow, Russia: Light and food industry (In Russian)
Sechin YT (Ed.). (1990). Methodological guidelines for assessing the abundance of fish in freshwater reservoirs. Moscow, Russia: VNIIPRKH (In Russian)
Sechin YuT (Ed.). (2010). Bioresource research on inland waters. Textbook. Kaluga, Russia: "Eidos" (In Russian)
Shashulovsky VA, Mosiyash SS (2005) Integrated assessment of the sustainability of the Volgograd reservoir ecosystem. Bulletin of Saratov State Agrarian University named after N.I. Vavilov 4:31–36 (In Russian)
Shashulovsky VA, Mosiyash SS, Ermolin VP, Anohina OK, Kostitsyn VG, MYu, Kudinov, Minin AE (2018) Features of the dynamics of commercial bioresources in the system of reservoirs of the Volga-Kama cascade. Fisheries 2:51–57 (In Russian)
Yudovich YuB, Dotsenko BN, Antonyuk AV (1982) Methodology for forecasting the catch of fish in lakes, rivers and reservoirs. VNIIPRKH, Moscow (In Russian)

Part II
Ecology and Protection of Aquatic Ecosystems

Chapter 4
Research Microplastics—Hydraulic Size of Microplastic Particles of Regular Shape and Their Distribution Over the Depth of the Watercourse

N. R. Akhmedova and V. A. Naumov

Abstract Features of calculating the settling velocity of microplastic particles (MP) of regular shape and their distribution over the watercourse depth are considered in the article. The balance equation of forces acting on the MP particle is solved numerically to determine the settling rate. In this case, the previously obtained dependence of the hydrodynamic drag coefficient on the Reynolds number and the particle shape is used. The method for constructing a profile of the MP distribution over the depth of a steady flow depending on the shape, size, and density of particles, and the average velocity of the watercourse is proposed. This profile allows you to calculate the average vertical concentration by measuring the MP concentration near the surface of the watercourse.

Keywords Microplastic · Hydraulic size · Settling rate

4.1 Introduction

The problem of environmental pollution by small particles (less than 5 mm) from polymer materials has become very relevant in the twenty-first century. These particles are called microplastics (MP). Initially, most research was carried out on the distribution of MP in the World's Oceans (see Andrady 2011; Chubarenko and Stepanova 2017; Cózar et al. 2014; Erni-Cassola et al. 2019). Gradually, it became clear that it is equally important to study the sources and ways of MP migration along rivers.

The contribution of a large urban area to MP pollution of the Los Angeles and San Gabriel rivers was evaluated in (Moore et al. 2011). Samples were obtained using nets from the mass ejection sites of MP. The nets made it possible to select particles larger than 1 mm. The number of plastic objects and fragments for 72 h of observations exceeded 2 billion with a total weight of more than 30 thousand kg. Bottom sediments

N. R. Akhmedova (✉) · V. A. Naumov
Kaliningrad State Technical University, Sovetsky prospect 1, 236022 Kaliningrad, Russian Federation
e-mail: isfendi@mail.ru

A. G. Arkhipov (ed.), *Sustainable Fisheries and Aquaculture: Challenges and Prospects for the Blue Bioeconomy*, Environmental Science and Engineering, https://doi.org/10.1007/978-3-031-08284-9_4

over 320 km of the Saint Lawrence river were studied in (Castañeda et al. 2014). Plastic beads were found in all samples. The average bead size was 0.7 mm in areas receiving municipal or industrial runoff. The concentration at different sites varied from 52 to 13,832 units per square meter of the bottom.

The results of the study of MP pollution on the surface of the Rhine River are presented in the article (Mani et al. 2015). Measurements were made at 11 points over 820 km. MP concentrations were varied along and across the river. The beads were predominant among the detected particles (58.5%). The authors (Mani et al. 2015) explain this fact by the widespread use of such particles in the manufacture of products made of polymer materials, as cleaning granules, and in other industries. Fragments of MP (37.5%) and fibers (2.5%) are formed as a result of destruction or abrasion of larger plastic products. Polystyrene was the dominant polymer among MP (29.7%).

The composition of MP in the surface waters of the lower Yellow River near the mouth was studied in (Han et al. 2020). Fibers accounted for more than 93%. Almost 88% of the particles in the river's surface waters were less than 0.2 mm. Infrared spectrometric analysis showed the following MP materials: polyethylene, polypropylene and polystyrene. The average MP content in the dry and wet seasons was 930 and 497 units per liter, respectively.

The aquaculture was the main source of MP pollution in the mouth of the Qin River (China) (Zhang et al. 2020). Surface water and sediment samples were collected using plankton nets at 12 points along the urban section of the river. Fibers made up the majority in surface waters. Particles of 1–5 mm in size predominated in the bottom sediments. Particles from polyethylene and polypropylene made up about 70%.

The results of studying MP at three stratified depths in the lower reaches of the Surabaya River (Indonesia) are presented in (Lestari et al. 2020). This river is a source of raw water, used for irrigation, for dumping household and industrial waste. More than 43 MP particles per cubic meter were in the surface level of the river. The MP content reached 34.6 units per m3 in the bottom layer. The MP content at medium depths was noticeably lower (0.76–12.6 units/m^3). Large MP films of 1–5 mm in size made of polyethylene and polypropylene prevailed at all depths.

These experimental studies show that MP particles of small size or density, as well as in the form of fibers or films, are concentrated near the surface of watercourses. MP particles of higher density and regular shape (beads, short cylinders) are located in the bottom layer of the river or are deposited on the bottom. The distribution of MP over the depth of the watercourse depends on the hydraulic particle size and characteristics of the river. The hydraulic particle size W_S is the rate of its steady-state deposition in still water.

The purpose of this work is to develop a method for constructing the distribution of MP in depth depending on the size, density, shape of particles and average flow velocity. MP fibers have their own characteristics (Naumov 2020a) and are not considered here.

4.2 Materials and Methods

4.2.1 *Velocity of Particles MP Steady Deposition*

The importance of correctly determining the WS value for the particle depth distribution is indicated in a number of papers (Dietrich 1982; Haider and Levenspiel 1989; Khatmullina and Isachenko 2017; Shraiber et al. 1990; Velikanov and Naumov 2014; Zhiyao et al. 2008). There are two main approaches.

The first approach is to use experimental data to obtain regression dependences of the deposition rate on the determining parameters. Dietrich (1982) used the dimensionless velocity W_*, the diameter D_*, and the Corey shape coefficient *csf*:

$$W_* = \frac{W_s^3}{G\nu},\ D_* = \frac{D_n^3 G}{\nu^2},\ G = g\frac{\rho_s - \rho_f}{\rho_f},\ csf = \frac{c}{(a\cdot b)^{0.5}} \tag{4.1}$$

where D_n is nominal (also called equivalent) particle diameter equal to the diameter of a sphere of the same volume as the particle; ν is the kinematic viscosity coefficient of the liquid; g is the acceleration of free fall; ρ_f, ρ_s are the density of fluid and particles, respectively; *a, b* and *c* are longest, middle and shortest axis of the particle in three orthogonal planes, respectively.

The second approach (Haider and Levenspiel 1989; Shraiber et al. 1990; Velikanov and Naumov 2014) is based on the fundamental laws of hydromechanics. The standard resistance curve of a sphere in a viscous medium is used in the second approach. The geometric shape coefficient is used to account for the shape of particles (Shraiber et al. 1990):

$$\xi = S/S_0,\ S_0 = (36\pi V^2)^{1/3} \tag{4.2}$$

where *S, V* are the surface area and volume of an arbitrary particle; S_0 is the surface area of a spherical particle of equal volume.

The sphericity coefficient $\phi = 1/\xi$ (Haider and Levenspiel 1989) is usually used in English-language publications.

The influence of differences in the particle shape from spherical coefficient of hydrodynamic resistance in the second approach takes into account the dynamic shape factor Γ.

Typically, the following formula is used:

$$C_\Gamma = \Gamma \cdot C_R(Re),\ = D_n \cdot W\nu \tag{4.3}$$

where C_R is the coefficient of hydrodynamic drag of a sphere in a boundless viscous medium; *Re* is the Reynolds number; *W* is the velocity of the particle relative to the liquid.

The coefficient Γ is a function of ξ and *Re* in the general case. The correction depends only on the geometric shape coefficient in the linear (*Re* < 1) and quadratic drag (*Re* > 1000) (Velikanov and Naumov 2014):

$$\Gamma_1(\xi) = 1 + 0.348 \cdot (\xi - 1), \ \Gamma_2(\xi) = 10.0 - 9.0/\xi \tag{4.4}$$

The transition drag region is the least explored. Moreover, the most deposition of MP occurs in this area. In our opinion, it is advisable to use the formula obtained in (Kondratev et al. 2016) from (4.4) in the transition drag region:

$$C_\Gamma = \left(\frac{24}{\mathrm{Re}} + \frac{53}{32 + \mathrm{Re}}\right) \cdot \Gamma_1(\xi) + 0.44 \cdot \Gamma_2(\xi) \tag{4.5}$$

Formula (4.5) is quite simple. The results of the calculation are consistent with the experimental data, at least as well as with the formulas of other authors.

The steady-state settling velocity of a dispersed particle can be found from a nonlinear equation. This equation does not have an analytical solution in the general case:

$$3\gamma \cdot D_n \cdot C_\Gamma\left(\frac{D_n W_s}{\nu}, \xi\right) \cdot W_s^2 = 4g(1 - \gamma), \ \gamma = \frac{\rho_f}{\rho_s} \tag{4.6}$$

Using the numerical method to find the WS from (4.6) could have caused difficulties 50 years ago. This problem is easily solved at present, for example, in the Mathcad (Naumov 2020b).

4.2.2 The Distribution of Particles MP Over the Watercourse Depth

The distribution of MP over depth depends on the action of gravity on suspended particles and the effect of turbulent water velocity pulsations. The long straight channel with a constant depth *H* and an average water velocity U_S is further considered. Large inertial particles are absent in this flow in the suspended state. They've all fallen to the bottom. The remaining small low-inertia particles form a background turbidity. They are in dynamic equilibrium. As many particles settle on the bottom, the same amount rises from the bottom. The average local turbidity of the watercourse does not change.

We assume that the MP impurity can be considered passive, and the velocity profiles of the carrier medium (water) are known. The distribution of each impurity fraction can be considered independently of the others. The vertical coordinate *Z* is counted from the bottom.

The change in the content of suspended MP particles over the flow depth is described by the turbulent diffusion equation. The diffusion equation can be simplified in a steady flow without particle deposition (Velikanov and Naumov 2014; Velikanov and Naumov 2017):

$$\frac{\partial}{\partial Z}\left(D_z\frac{\partial B_i}{\partial Z}\right)+W_i\frac{\partial B_i}{\partial Z}=0 \tag{4.7}$$

where D_z is the coefficient of vertical turbulent diffusion; B_i is the concentration of the i-th fraction of MP; W_i is the settling rate of the i-th fraction (hydraulic fineness).

We can write for a power model of the longitudinal velocity profile of a watercourse U (Volynov 2013):

$$U=U_{max}\cdot z^{nu},\ z=\frac{Z}{H},\ \frac{U_{max}}{U_s}=nu+1,\ nu=\frac{\sqrt{g}}{0.4Cs} \tag{4.8}$$

where nu is the exponent in the vertical velocity profile; Cs is the Chezy coefficient.

The vertical turbulent diffusion coefficient, which varies in depth, is calculated using the formula from (Volynov 2013):

$$D_z=\frac{g\cdot H\cdot U_s(1-z)}{(1+nu)\cdot Cs^2\cdot nu\cdot z^{nu-1}} \tag{4.9}$$

The formula for the profile of the dimensionless impurity concentration MP is obtained by substituting (4.9) in (4.7) and integrating:

$$b_i(z)=\frac{B_i(z)}{B_i(0)}=exp\left(-\frac{(nu+1)\cdot W_{si}}{0.16\cdot nu\cdot U_s}\cdot\int\limits_0^z\frac{d\eta}{\eta^{1-nu}(1-\eta)}\right) \tag{4.10}$$

where $B_i(0)$ is the concentration of the i-th MP fraction at the bottom.

4.3 Results

We calculate the steady-state settling velocity under experimental conditions (Khatmullina and Isachenko 2017) using formulas (4.4)–(4.6). Polycaprolactone (PCL) density ρS = 1131 kg/m^3; cylinder diameter range D = 0.59–5.09 mm; cylinder elongation λ ≡ L/D = 0.6–1.26; kinematic water viscosity coefficient ν = 0.872·10–6 m^2/s. Let's express the geometric shape coefficient and the equivalent cylinder diameter in terms of elongation:

$$\xi = \frac{\lambda + 0.5}{(1.5 \cdot \lambda)^{2/3}}, \quad D_n = D \cdot (1.5 \cdot \lambda)^{1/3} \tag{4.11}$$

The results of calculating the settling velocity of cylinders with two elongation values of 0.6 and 1.26 are shown in Fig. 4.1.

The results of calculating the MP distribution over the watercourse depth using the formula (4.10) are shown in Figs. 4.2, 4.3, 4.4, 4.5. In all cases, $b_i(0) = 1$ (the lower right point of the graphs).

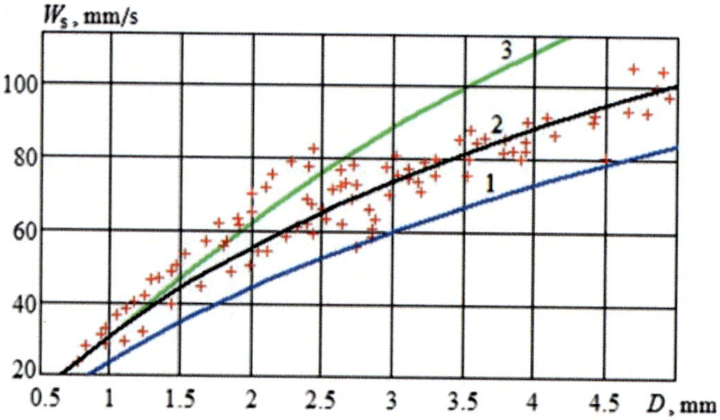

Fig. 4.1 Dependence of the settling velocity of the PCL cylinders on the diameter. Points are experimental data (Khatmullina and Isachenko 2017). Curves are calculation results for (3)–(5): 1—cylinders at $\lambda = 0.6$; 2—cylinders at $\lambda = 1.26$; 3—deal spheres (compiled by the authors)

Fig. 4.2 Effect of the MP size on the distribution ($\rho_S =$ 1131 kg/m^3; $\xi = 1.6$; $U_{S^*} =$ 0.15 m/s): 1 – $D_n =$ 0.02 mm; 2 – $D_n =$ 0.045 mm; 3 – $D_n =$ 0.070 mm; 4 – $D_n =$ 0.10 mm (compiled by the authors)

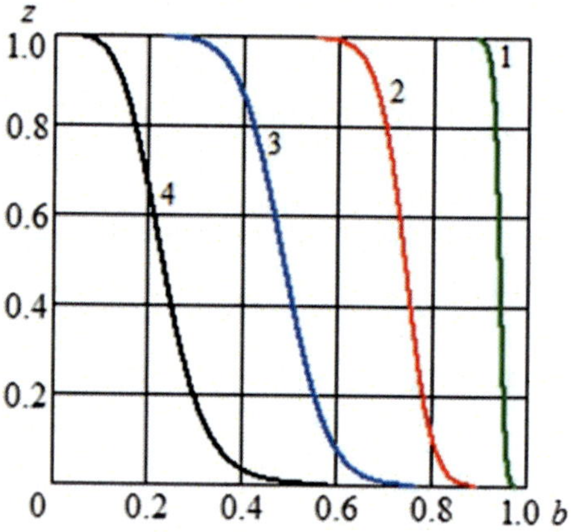

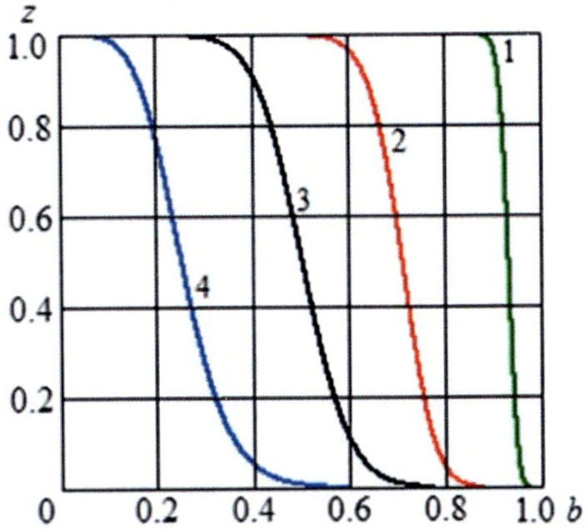

Fig. 4.3 Effect of MP density on the distribution ($D_n = 0.045$ mm; $\xi = 1.6$; $U_{S\cdot} = 0.15$ m/s): 1 – $\rho_S = 1030$ kg/m^3; 2 – $\rho_S = 1150$ kg/m^3; 3 – $\rho_S = 1300$ kg/m^3; 4 – $\rho_S = 1600$ kg/m^3 (compiled by the authors)

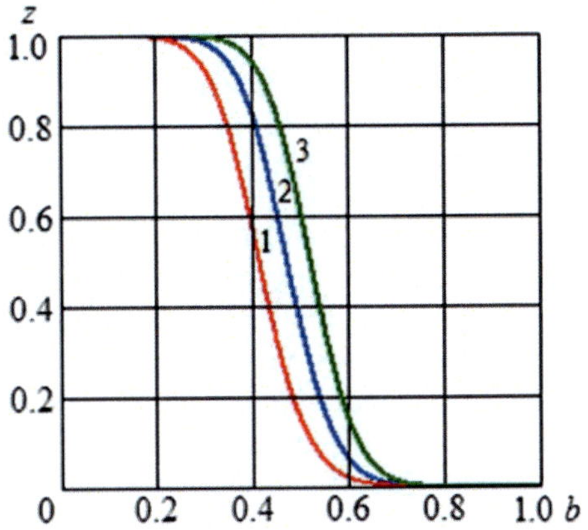

Fig. 4.4 Effect of the MP shape coefficient on the distribution ($D_n = 0.07$ mm; $\rho_S = 1131$ kg/m^3; $U_{S\cdot} = 0.15$ m/s): 1 – $\xi = 1.0$; 2 – $\xi = 1.5$; 3 – $\xi = 2.0$ (compiled by the authors)

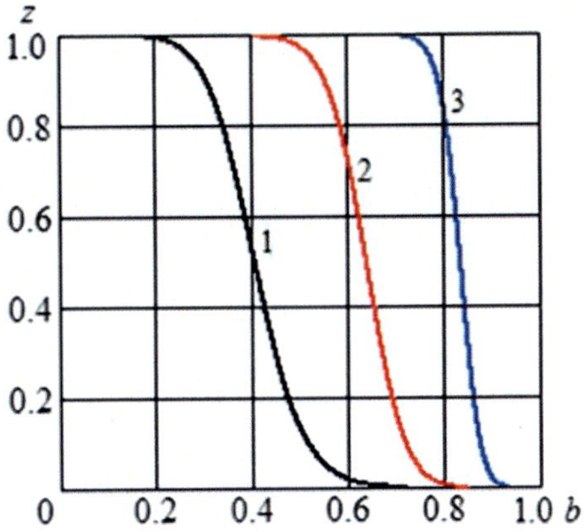

Fig. 4.5 Effect of water flow velocity on the distribution ($D_n = 0.045$ mm; $\rho_S = 1131$ kg/m^3; $\xi = 1.6$): 1 – $U_{S\cdot} = 0.05$ m/s; 2 – $U_{S\cdot} = 0.1$ m/s; 3 – $U_{S\cdot} = 0.3$ m/s (compiled by the authors)

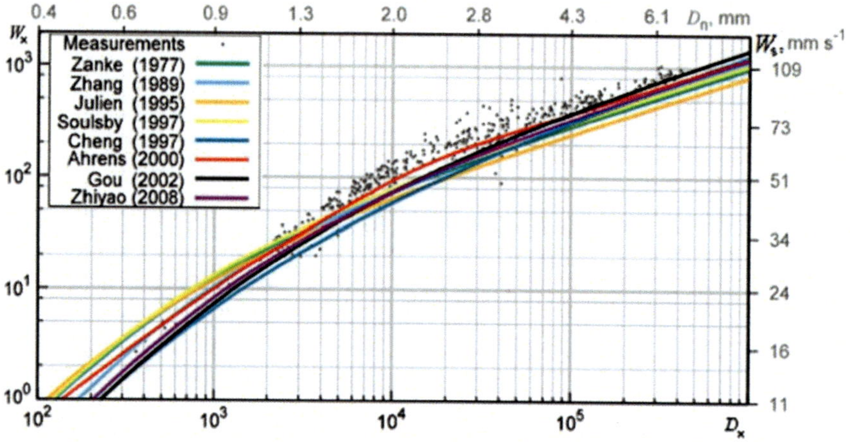

Fig. 4.6 Dimensionless settling velocity of polycaprolactone cylinders (points are experimental data) and calculated curves. Figure taken from (Khatmullina and Isachenko 2017)

4.4 Discussion

Quite complex regression dependences of W* on D* and csf were obtained in (Dietrich 1982). Obtaining new experimental data in the first approach requires selecting new empirical coefficients in a known relationship or proposing a new regression function. This approach is not fundamental and does not allow us to identify errors and experimental errors. A significant part of the experimental data in Fig. 4.6 from (Khatmullina and Isachenko 2017) is located above the calculated curves of various authors from the review (Zhiyao et al. 2008).

All the samples in the experiments (Khatmullina and Isachenko 2017) had elongation values of 0.6 and 1.26. Therefore, the experimental points in Fig. 4.1 should lie between curves 1 and 2. But many of them lie above curve 2. Moreover, some of the points lie even above curve 3, which is constructed for ideal spheres. The result was an overestimation of the settling velocity of cylindrical MP particles in (Khatmullina and Isachenko 2017). Most likely, there was a systematic error.

4.5 Conclusion

For MP particles with a density of 1131 kg/m^3 and an equivalent diameter of less than 20 microns, the vertical distribution in the watercourse differs little from uniform. With increasing D_n, the unevenness of the profile increases. The results of the study showed that a significant part of the MP with a high density moves near the bottom, even if the particles are small in size, and a decrease in the velocity of the watercourse leads to an increase in the uneven distribution of MP on the vertical. The geometric coefficient of the particle shape with a change from 1 to 2 slightly affects the distribution of MP in depth. As a rule, when assessing the contamination of a water body with microplastics, they are limited to sampling from the surface layer, which does not provide complete information about its condition. The method proposed by the authors for constructing the distribution of MP by depth, depending on the shape, size and density of particles, as well as the average flow velocity, will allow us to estimate the average concentration of MP measured in the surface layer over the entire depth of the flow (vertically). This method can be used when conducting environmental studies on watercourses.

References

Andrady AL (2011) Microplastics in the marine environment. Mar Pollut Bull 62:1596–1605

Castañeda RA, Avlijas S, Simard MA, Ricciardi A (2014) Microplastic pollution in St. Lawrence river sediments. Canadian J. Fisheries and Aquatic Sciences 71:1767–1771

Chubarenko I, Stepanova N (2017) Microplastics in sea coastal zone: lessons learned from the Baltic amber. Environ Pollut 224:243–254. https://doi.org/10.1016/j.envpol.2017.01.085

Cózar A, Echevarría F, González-Gordillo JI, Irigoien X, Ubeda B, Hernández-León S, Palma AT, Duarte CM (2014) Plastic debris in the open ocean. Proc Natl Acad Sci 111(28):10239–10244

Dietrich WE (1982) Settling velocity of natural particles. Water Resour Res 18:1615–1626

Erni-Cassola G, Zadjelovic V, Matthew I, Gibson MI, Christie-Oleza JA (2019) Distribution of plastic polymer types in the marine environment. J. Hazardous Materials 369:691–698

Haider A, Levenspiel O (1989) Drag coefficient and therminal velocity of spherical and nonspherical particles. Powder Technol 58:63–70

Han M, Niu X, Tang M, Zhang B, Wang G, Yue W, Kong W, Zhu J (2020) Distribution of microplastics in surface water of the lower Yellow River near estuary. Sci Total Environ 707:135601

Khatmullina L, Isachenko I (2017) Settling velocity of microplastic particles of regular shapes. Mar Pollut Bull 114:871–880

Kondratev AS, Nya TL, Shvydko PP (2016) Calculation of hydraulic drag coefficient of arbitrary shape solid particles. Fundamental Research 11:286–292

Lestari P, Trihadiningrum Y, Wijaya BA, Yunus KA, Firdaus M (2020) Distribution of microplastics in Surabaya River. Indonesia. Science Total Environment 726:138560

Mani T, Hauk A, Walter U, Burkhardt-Holm P (2015) Microplastics profile along the Rhine River. Sci Rep 5:179–188

Naumov VA (2020a) Law of distribution of breaking elongation of yarn. Fibre Chem 51:244–246

Naumov, V. A. (2020b). Influence of leakage on characteristics of the vacuum transport unit based on the water-ring pump. IOP Conf. Series: Materials Science and Engineering, 862, 032007.

Moore CJ, Lattin GL, Zellers AF (2011) Quantity and type of plastic debris flowing from two urban rivers to coastal waters and beaches of Southern California. J. Integrated Coastal Management 11:65–73

Shraiber AA, Gavin LB, Naumov VA, Yatsenko VP (1990) Turbulent Flows in Gas Suspensions. Hemisphere, New York, p 248

Velikanov NL, Naumov VA (2014) Suspended particles in a water stream and their deposition. Water: chemistry and ecology, 2, 114–119. (In Russian)

Velikanov NL, Naumov VA (2017) Modeling of the distribution of suspended organic impurities in watercourses. Water: chemistry and ecology, 3, 3–8 (In Russian)

Volynov MA (2013) Influence of the planned geometry of the riverbed on the diffusion and dispersion of impurity. Fundamental Research 6:535–540

Zhang L, Liu J, Xie Y, Zhong S, Yang B, Lu D, Zhong Q (2020) Distribution of microplastics in surface water and sediments of Qin river in Beibu Gulf, China. Sci Total Environ 708:135–176
Zhiyao S, Tingting W, Fumin X, Ruijie L (2008) A simple formula for predicting settling velocity of sediment particles. Water Science and Engineering 1:37–43

Chapter 5
Assessment of the Trophic Status of the Coastal Area of Lake Vistytis (Kaliningrad Region) by Hydrochemical Parameters

E. V. Krivopuskova and N. N. Tzvetkova

Abstract The coastal areas of water bodies have always been a place of concentrated recreational infrastructure, and therefore they were subjected to significant anthropogenic load. However, not only terrestrial ecosystems were adversely affected, economic activities within the watershed area and heavy recreation load have influence on coastal waters. Coastal waters of freshwater bodies have been exposed to the strongest anthropogenic load in the past few years as a result of the overloaded sea coasts and increased population mobility. The aim of this study was to analyze the changes in the trophic status in the coastal part of Lake Vistytis based on long-term data on total phosphorus and chlorophyll "a" concentrations. In Kaliningrad region, Lake Vistytis is a habitat for several valuable commercial fish populations, and its shallow water ensures their recruitment. The obtained data show that over the past few years the concentration values of the analyzed indicators in summer are sufficiently high, which is indirectly confirmed by the abundant algal blooms found throughout the coastal waters. Compared with the initial study period (2014–2016), when the trophic index values fell between the oligotrophic and mesotrophic status, it is currently (2019–2020) observed to grow, particularly in summer. In previous years, relatively high values of the index were associated with temperature conditions during the growing season, in recent years however, the index has been strongly related to the level of recreational load.

Keywords Trophic index · Ecological status · Anthropogenic impact · Phosphorus · Chlorophyll "a" · Shallow waters

E. V. Krivopuskova (✉) · N. N. Tzvetkova
Kaliningrad State Technical University, Sovetsky prospect 1, 236022 Kaliningrad, Russian Federation
e-mail: katekrivopuskova@gmail.com

N. N. Tzvetkova
e-mail: nagornova@klgtu.ru

A. G. Arkhipov (ed.), *Sustainable Fisheries and Aquaculture: Challenges and Prospects for the Blue Bioeconomy*, Environmental Science and Engineering, https://doi.org/10.1007/978-3-031-08284-9_5

5.1 Introduction

At present, the development of transport infrastructure and the economy of urbanized areas make unique natural ecosystems accessible to a large number of people, thereby causing regular changes, which sometimes lead to catastrophic aftereffects. An increased flow of tourists increases the human impact on the ecosystem, which results in deteriorated soil quality, disturbed grass cover, decreased species diversity and degraded quality of water bodies (Rossolimo 1977). Tourism is one of the main types of the economic activity that has been actively developing in recent decades, and it has a strong impact on natural ecosystems.

Modern tourism development programs are increasingly implementing the principles of green economy to fully exploit the potential of natural complexes with the minimum possible anthropogenic impact on them. The quality of both natural and anthropogenic environment is crucial for tourism development. The relationship between the development of recreational infrastructure and the state of the environment is complicated, since many types of activities can have an adverse impact on the natural ecosystem and decrease its recreational attractiveness and deteriorate its bioproductivity (in terms of qualitative and quantitative indicators).

Among all natural ecosystems, the most significant for recreational development are the coastal areas of water bodies, which in turn are considered the basic unit for the development of the blue economy, aimed to fully exploit the water ecosystem potential. In the coastal area of the water bodies, recreation centers are being built, camping sites are being organized, berths for water transport are being equipped, and beaches are being created. Increasing rates of urbanization of the coastal areas increase the input of organic matters and nutrients in water bodies, which in turn changes hydrological and hydrochemical regimes of the ecosystem, and increase its trophic status.

The coastal areas of water bodies respond particularly sharply to the increased anthropogenic load. In these areas, "excessive" volumes of nutrients and organic matters, coming from the watershed area and formed in the water body accumulate faster compared to other areas. Yet, these are the areas where the food base for aquatic organisms is formed, spawning grounds for most commercial fish species are located, and the recruitment of fish populations at early developmental stages occurs.

The coastal area of lakes receives a large amount of light, quickly responds to changes in temperature, oxygen and ice conditions, and water transparency. Change in the water level, both seasonal and daily, causes changes in the area and configuration of the littoral zone, and, hence, the concentration of nutrients inflowing from the watershed area (Stepanova et al. 2020).

Therefore, assessment of the trophic status of water bodies is an important stage in the study of aquatic ecosystems which helps to elaborate programs for coastal area development, and to develop measures for their conservation. According to Naumann (Naumann 1929), the trophic status of the water body is defined as the biotic response of aquatic ecosystems to the influx of organic matters and nutrient. Thus, a comprehensive characteristic of changes in the aquatic ecosystem trophic status requires extensive data on its abiotic and biotic states, knowledge of the relationships between hydrochemical, hydrobiological and ichthyological parameters, and information on the sources and level of the anthropogenic impact on the water body.

At the same time, these data require lengthy primary processing, and often necessitate selection of one or more groups of parameters. Most often, the trophic status is assessed by hydrochemical parameters, such as the concentration of total phosphorus and total nitrogen, water transparency, the concentration of chlorophyll "a", the change in the pH, the value of biochemical oxygen consumption, and the oxygen regime of the water body.

Hydrochemical parameters of the aquatic environment provide a sufficiently large amount of data to assess the trophic status of the water body and changes occurring in it under anthropogenic impact within a short time period and with minimal material processing costs.

In the last decade, the Kaliningrad region has been actively implementing programs aimed to increase the tourist attractiveness of the region, with new recreational objects being created within their framework. One of these objects is the natural park "Vishtynetsky" founded in 2010 with a total area of 22 thousand hectares and including Lake Vistytis, the largest lake in the region. The program for the development of this territory provides for the recreational development of the coastal area and the water area of the lake with regard to the ecological and legislative status of the natural park (Resolution of the Government of the Kaliningrad region 2012).

Since the time the tourism development strategy was initiated in the Kaliningrad region and pandemic restrictions were introduced, an organized and unorganized tourism flow was observed to annually regularly increase (approximately by up to 40% annually) within the coastal area of Lake Vistytis. Tentative calculations show that in total, during the recreational season (from May to August), taking into account weather conditions (air temperature above 22 °C and lack of rainy weather), up to 40 thousand people on average came to the area (398 people/day) (Krivopuskova and Sokolov 2016). At the same time, the concentration of nutrient and organic matters in the coastal area of the lake tends to increase.

One of the indicators of the increased anthropogenic load on Lake Vistytis is increasing water bloom, though in the early 2010s, rapid development of phytoplankton was observed in the Lithuanian coastal area. Currently, water bloom can be observed throughout the water area. The strongest adverse impact of the rapid development of phytoplankton can be found in the northern part of the lake, and particularly in the area of the headwater of the Pissa River, since this is the only place that ensures the flowage of the lake.

Lake Vistytis (see Fig. 5.1) is a transboundary water body located in the southeast of the Kaliningrad region. This is the only large freshwater body in the territory of the Kaliningrad region, which is part of the Masurian lake district. It is located on the edge of the Vishtynetsko-Suwalki upland at an altitude of 170 m a.s.l. The Russian-Lithuanian border passes through the water area of the lake.

The watershed area of the lake, with a total area of 281.9 km^2 (Krivopuskova and Sokolov 2016), is located in the territory of three countries, which determines

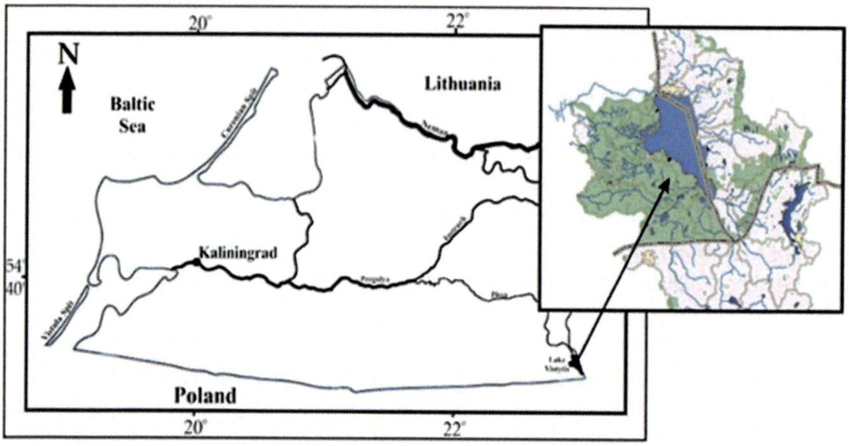

Fig. 5.1 Location of Lake Vistytis in the Kaliningrad region (left) and its detailed map (right) (*Source* compiled by the authors)

the diversity of sources of nutrients entering the lake. Active use of coastal areas for purposes increased the number of recreation centers. In the last few years, the recreational load on the coastal territories of the Russian part of the watershed area has increased, which in turn has an impact on the state of the coastal ecosystem of Lake Vistytis.

The main aim of this study was to assess the ecological status of the coastal waters of the Russian part of Lake Vistytis during the period of increased anthropogenic load.

5.2 Materials and Methods

A set of stations (see Fig. 5.2) was formed with respect to the hydrological and morphological features of the lake, and the levels of recreational development of the territory, as the main source of "additional" organic matter and nutrients.

The stations are grouped with respect to the level of anthropogenic load as follows:

- southern (18a, 18, 17, 20) stations in this area of the lake exhibit low anthropogenic impact, low level of development of higher aquatic vegetation. In this area, the main inflow of water from the watershed area occurs through the Chernitsa River;
- central (42, 4b and 68) stations with the strongest and most continuous anthropogenic impact;
- Utinyy Bay (5) is a vast shallow area with a high trophic status (mesotrophic) (Bernikova 2013) compared to other areas and a high percentage of overgrowth of the coastal areas with higher aquatic vegetation;

Fig. 5.2 Map of water sampling (numbers in red), and bathymetry (numbers in bold) of Lake Vistytis *Source* compiled by the authors

- northern (6, 21 and 11) stations with increased anthropogenic load due to the development of infrastructure and recreational facilities, the area, where the only runoff from the lake (headwater of the Pissa River) is located.

Samples were taken in the coastal area of the lake (see Fig. 5.2) at a depth of 1.5 to 2.0 m. The studies employed data collected in summer 2014–2020 as part of the comprehensive environmental monitoring conducted by the Department of Ichthyology and Ecology, KSTU. Simultaneously with sampling for hydrochemical analysis, water temperature, electrical conductivity, pH and dissolved oxygen concentration were measured in situ.

The water samples were analyzed using standard methods (Baird et al. 2017) in laboratory conditions. The chlorophyll "a" concentration was determined by the spectrophotometric method using 90% acetone solution.

For quick and reliable assessment of the trophic status of the coastal area of the lake, the Carlson trophic index (Carlson 1977) was used, which showed the greatest efficiency and applicability for a large number of aquatic ecosystems on a global scale. The index is calculated through logarithmic relationship between the trophic status of the water body and such parameters as water transparency, chlorophyll "a" concentration, and concentrations of total nitrogen and total phosphorus. The trophic status can be assessed by one parameter and as combination of parameters, but the more parameters are used, the more accurate and reliable assessment of the status of the aquatic ecosystem.

The Carlson's trophic index is recommended as the simplest method for calculating the trophic status. A widespread use of this index and an increased number of calculation parameters used show that the indices calculated for chlorophyll "a" and total phosphorus are more indicative and can be used to assess water trophy not only in the aquatic ecosystem as a whole, but also in some individual areas, particularly coastal ones, which are primarily exposed to anthropogenic impact.

The Carlson's trophic index is calculated using chlorophyll "a" and total phosphorus concentrations by the formulas (Carlson 1977):

$$TSI\,(CHL) = 9.81\,\ln(CHL) + 30.6 \tag{5.1}$$

$$TSI\,(TP) = 14.42\,\ln(TP) + 4.15 \tag{5.2}$$

To assess changes in the trophic status of the coastal area of Lake Vistytis, long-term data on chlorophyll "a" and total phosphorus concentrations were employed as most indicative, with due regard to changes in the coastal area.

5.3 Results

The ecosystem of Lake Vistytis accumulates all forms of phosphorus, both phosphorus coming from outside and that produced inside the lake. Phosphorus is one of the biogenic elements that affect the rate of production in the water body (Rossolimo 1977). The total phosphorus concentration in water depends on different factors, such as weather conditions during the growing season, water temperature, and the intensity of human activity. Therefore, only long-term data collected in similar periods can provide full picture on the ecological status of the water body.

The total phosphorus concentration in the coastal area of the lake varied greatly during the study period (Table 5.1). Yet some patterns were revealed. Over the studying period, a significant amount of phosphorus was found in Utinyy Bay at station 5, an area with a high trophic status (recorded since the early 1980s (Bernikova et al. 2013)) and low water circulations. Inflows with higher phosphorus concentration could be clearly visible from the coastal area of the lake to the deep water. The latter were most distinctly observed in 2016, when the inflow with the maximum phosphorus concentration (0.087 mg/l at station 15) flew from the Lesnoy stream

Table 5.1 Concentrations of total phosphorus, chlorophyll "a" and the number of tourists in the coastal area of Lake Vistytis during the study period

	Years						
	2014	2015	2016	2017	2018	2019	2020
Concentration of total phosphorus, mg/l *	0,06	0,05	0,07	0,08	0,31	0,56	0,96
Chlorophyll "a" concentration, mg/l *	1,89	1,94	2,69	2,75	2,71	2,72	3,12
Number of tourists **	410	400	470	720	890	765	1240

Note *—average value in the coastal area; **—the number of tourists was estimated visually during the period of work running through the campsite located on the east-southeast coast. The total phosphorus concentration tends to increase in the coastal area of the lake, particularly in areas crowded with tourists.

Chlorophyll "a" is the major phytoplankton pigment actively involved in photosynthesis and a unique natural marker that reflects internal processes occurring in water bodies, such as formation, migration and transformation of organic matter (Bernikova et al. 2015). Data on qualitative and quantitative characteristics of photosynthetic pigments illustrate the mechanism of interaction between production and destruction processes in different types of water bodies and can be used to forecast aftereffects of the anthropogenic impact on aquatic ecosystems. One of the factors increasing the production rate in water bodies is the supply of biogenic elements, in particular phosphorus. In the study period, the chlorophyll "a" concentration varied from 0.441 to 6.740 mg/l. Chlorophyll "a" distribution in the lake as the same as the total phosphorus concentration vary greatly in different years. Yet the chlorophyll "a" concentration tends to increase similar to that of total phosphorus.

In addition, an increased accumulation of large representatives of blue-green algae is observed in the water column within the photic layer. Thus, in some periods, high concentrations of algae could be observed at a depth of up to 7–8 m (the thermocline was located at a depth of 12 m), which was later confirmed by the analysis of chlorophyll "a" concentrations. During these periods, the chlorophyll "a" concentration in the coastal area is much higher than the seasonal average averages, which affects water color.

It should be noted that there is no evident relationship between the chlorophyll "a" concentration and potential sources of pollution with nutrients, such as places of organized recreation in the north-eastern part of the lake adjacent to the village Vishtitis due to a strong effect of the wind-induced water current. Similar distribution patterns were observed for total phosphorus.

The dependence of the studied total phosphorus and chlorophyll "a" concentrations (see Fig. 5.3) is an important indicator of not only production processes in water bodies, but also a necessary parameter for creating functional models of the

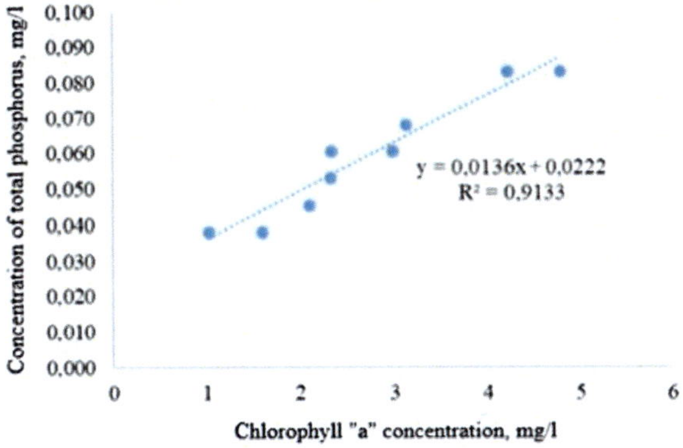

Fig. 5.3 Correlation dependence between the concentrations of total phosphorus and chlorophyll "a" based on the data obtained over the period of 2014–2020 *Source* compiled by the authors

ecosystem of Lake Vistytis. The data obtained showed a directly proportional relationship between the considered parameters not only in the coastal area of the lake, but also in its deep water areas.

Analysis of the literature data showed that production processes in Lake Vistytis are similar to those in the lakes of the north-west of Russia, which are in a transitional state from oligo- to mesotrophic trophic status (Kitaev 2007). Preliminary data on the relationship between the concentrations of total phosphorus and chlorophyll "a" can make the basis for creation of the aquatic ecosystem production models and for development of programs for monitoring of the state of Lake Vistytis with regard to the level of the anthropogenic impact.

5.4 Discussion

The data on total phosphorus and chlorophyll "a" concentrations in the coastal waters of Lake Vistytis revealed high concentration values in the summer period over the past few years, which is indirectly evidenced by abundant algal blooms throughout the coastal waters. This was particularly evident during the "pandemic" period of study, when the recreational load increased significantly. As a result, the Carlson's trophic index calculated using data obtained in recent years is significantly high (62) and indicates an excess water trophy in summer. Compared to the initial study period (2014–2016), when the index values were borderline between the oligotrophic and mesotrophic status, the index is observed to increase in summer. In previous years, relatively high values of the index were associated with temperature conditions during the growing season. In recent years, there has been a strong relationship between the index value and the level of recreational load.

Over the past years, the index value indicates a eutrophic status of the coastal area of the lake in summer. However, the study results obtained in other seasons show that total phosphorus is actively involved in the circulation, and the phyto- and bacterioplankton biomass accumulated due to its consumption virtually completely transforms at higher levels of the food chain. Excess biomass also gradually accumulates in the form of dead organic matter in bottom sediments.

The highest indicator of the Carlson's trophic index was found in 2018 (84.8), and the lowest one was observed in 2014 (39.7). It should be noted that in summer 2018 and 2014 the water temperature was similar and sufficiently low compared to the long-term averages, which affected the use of inorganic forms of phosphorus by phytoplankton.

However, the values of the Carlson's trophic index calculated using data on the chlorophyll "a" concentration were more indicative and relatively correlated with the trophic status data assessed by other indicators (Krivopuskova et al. 2017; Masyutkina 2014). The chlorophyll "a" concentration indirectly indicates an approximate phytoplankton biomass. There is a natural trend, the more chlorophyll, the more phytoplankton and the higher the water trophy in the lake. Hosmani (Hosmani 2010) suggests that chlorophyll "a" concentration values can be used as the main indicators for assessing the trophic status of the lake and the processes occurring in it.

According our researches for Lake Vistytis, the chlorophyll "a" concentration was most indicative for assessing water trophy, and a significant increase in phytoplankton biomass was recorded not only in laboratory studies, but also in field observations. Since 2014, summer algal "bloom" has been observed in amounts not typical of the previous periods with a tendency of increased duration and intensity.

The average value of the Carlson's trophic index calculated based on the chlorophyll "a" concentration (40.7) in the study period indicates coastal waters of Lake Vistytis as mesotrophic. In 2018 and 2019, the index is slightly lower and indicates a transitional status (between oligotrophic and mesotrophic). Low values of the index correspond to high concentrations of phosphorus compounds; for a number of reasons, mainly due to the features of the temperature regime, phosphorus has not yet been fully used for phytoplankton biomass formation.

In general, in favorable conditions and in the absence of weather anomalies, chlorophyll "a" concentrations in coastal waters of the lake can be used to assess their trophic status by the Carlson index as mesotrophic. In previous years (Bernikova 2013), these areas were assessed as oligorophic with some features of mesotrophy.

5.5 Conclusion

Assessment of water trophy shows that its levels are only stages between the oligotrophic and eutrophic status in a continuous chain of trophic transformations. A lake can be eutrophic according to some indicators while retaining the features of mesotrophy and even oligotrophy according to others.

Trophic status monitoring is an important part of assessment and management of water ecosystems. Since the phosphorus concentration is a limiting factor for algae growth (Rossolimo 1977), its value is used to assess water trophy. Yet the obtained values are not always reliable, similar to our case.

The formation of algae biomass is affected by temperature conditions, which can significantly limit the use of inorganic phosphorus for algae growth. Thus, the values of trophic indices calculated using only one parameter can be overestimated, which was found in our study as well.

Therefore, the Carlson's trophic index calculated based on the chlorophyll "a" concentration is most appropriate for assessing the trophic status of Lake Vistytis with due regard to ongoing changes within its watershed and coastal area.

To minimize an adverse impact of the increasing flow of tourists on the ecosystem of Lake Vistytis, it is necessary to control the number of tourists and organize centralized, equipped and accessible recreation areas. Further development of the recreational sector within the coastal area of the lake should be based on principles of rational and sustainable development specified in blue economy programs.

The study yielded the following conclusions:

(1) The analysis of the study results revealed a steady increase in total phosphorus and chlorophyll "a" concentrations in the coastal area of Lake Vistytis in the growing season, which is not related to weather conditions.
(2) The main cause of the increased concentration of biogenic elements in the coastal areas of the lake in the warm season is a growing number of tourists and increased anthropogenic load on the watershed area. Due to insufficiently equipped recreation areas, surface washoff contributes to penetration of the additional amount of total phosphorus into the lake from the watershed area.
(3) The value of the Carlson's trophic index in some years (for example, 2018) indicates the eutrophic status of the coastal area. Earlier, the trophic status of Lake Vistytis in the coastal area was assessed as oligotrophic with features of mesotrophy, and currently the trophic status of the coastal area of the lake is assessed as mesotrophic, therefore, it can be argued that the ecological status of the lake in the coastal area is deteriorating.
(4) To preserve the ecosystem of Lake Vistytis and its attractiveness for tourism and recreation, a comprehensive environmental monitoring of its status should be performed not only in the growing season, but also after the tourist season. This will enable monitoring of the accumulation of mineral forms of nitrogen and phosphorus and indicate the trends in the trophic status of Lake Vistytis.

References

Baird RB, Bridgewater LL, Eaton AD, Rice EW (eds) (2017) Standard methods for the examination of water and wastewater, 23rd edn. APHA, New York

Bernikova TA, Nagornova NN, Tsupikova NA (2013) Long-term changes in water consumption of the population as an indicator of the trophic level of Lake Vistytis. KSTU NEWS 28:70–78 (in Russian)

Bernikova TA, Krivopuskova EV, Tsvetkova NN (2015) Chlorophyll "a" as indicator of the trophic status of Lake Vistytis. In: Aquatic bioresources, aquaculture and ecology of water bodies: Proceedings of the III Baltic Marine Forum. KSTU, Kaliningrad, pp. 151–153 (in Russian)

Carlson RE (1977) A trophic state index for lakes. Limnol Oceanogr 22(2):361–369. https://doi.org/10.4319/lo.1977.22.2.0361

Hosmani SP (2010) Trophic state index in conservation of lake ecosystems. Advances in Plant Sciences 23(II): 593–596

Kitaev SP (2007) Basic general limnology for hydrobiologists and ichthyologists. Karelian scientific. Center of the Russian Academy of Sciences, Petrozavodsk (in Russian)

Krivopuskova EV, Bernikova TA, Agarkova TS (2017) Assessment of the ecological state of Lake Vistytis (Kaliningrad region) according to hydrochemical indicators. In: Aquatic bioresources, aquaculture and ecology of water bodies: proceedings of the V Baltic Marine Forum. KSTU, Kaliningrad, pp. 192–197 (in Russian)

Krivopuskova EV, Sokolov AV (2016) Sources of anthropogenic load on Lake Vistytis within its catchment area. In: Aquatic bioresources, aquaculture and ecology of water bodies: Proceedings of the IV Baltic Marine Forum. KSTU, Kaliningrad, pp. 43–46 (in Russian)

Masyutkina EA (2014) Assessment of the ecological state of the Lake Vistytis with the use of various hydrobiological indices. Bulletin of the Baltic Federal University. I. Kant 7:66–76 (in Russian)

Naumann E (1929) The scope of chief problems of regional limnology. Int.ReviewGes.Hydrobiol. 21: 423–444. https://doi.org/10.1002/iroh.19290220128

Resolution of the Government of the Kaliningrad region (2012) On the organization of the natural park "Vishtynetsky" (January 19, 2012 No 9) Kaliningrad, Russia (in Russian)

Rossolimo LL (1977) Changes in limnic ecosystems under the influence of the anthropogenic factor. Nauka, Moscow (in Russian)

Stepanova AB, EYu, Voyakina, Babin AV, Zueva NV, Zuev YuA (2020) Results of studies of the coastal zone of Lake Ladoga in the area of Valaam Archipelago (1998–2019) at RSHU. Proceedings of the Russian State Hydrometeorological University 60:325–349. https://doi.org/10.33933/2074-2762-2020-60-325-350(inRussian)

Chapter 6
Vistula Lagoon as Potential Object of Accumulated Environmental Harm

A. V. Minashkina and S. V. Kondratenko

Abstract The initial list of HELCOM hot spots in the Kaliningrad Region included 10 objects. There are now 7 objects left with the Vistula Lagoon among them. Numerous rivers of the Kaliningrad Region enter the bay, including the largest Pregolya River. Until 2019, the largest landfill that lacked a system for protecting the aquatic environment from the resulting leachate was located on the coast of the bay. Different chemical pollutants entered the Vistula Lagoon and accumulated in water, bottom sediments and hydrobionts, posing a threat to the ecosystem of the Vistula Lagoon, to residents of the Kaliningrad Region and numerous tourists. Detailed long-term monitoring of the content of nutrients in the water of the bay has been performed, yet the chemical pollution of the bay is studied poorly. The data on the content of organic and inorganic substances in fish and bottom sediments have become available only recently. It was established that pollutants that entered numerous landfills could enter the soil and then, through dips in impermeable soils, aquifers that discharge (release) into the Vistula Lagoon. In this regard, it is proposed to give it the status of an "object of accumulated ecological harm", the definition of which recently appeared in the Federal Law "On Environmental Protection". This status will allow a more detailed study of the bay in terms of its environmental safety.

Keywords Vistula Lagoon · Chemical pollution · Ecological monitoring · Heavy metals · Object of accumulated ecological harm

A. V. Minashkina · S. V. Kondratenko (✉)
Kaliningrad State Technical University, Sovetsky prospect 1, 236022 Kaliningrad, Russian Federation
e-mail: kondrat@klgtu.ru

A. G. Arkhipov (ed.), *Sustainable Fisheries and Aquaculture: Challenges and Prospects for the Blue Bioeconomy*, Environmental Science and Engineering, https://doi.org/10.1007/978-3-031-08284-9_6

Fig. 6.1 Location of the Vistula Lagoon. *Source* yandex.ru/maps/

6.1 Introduction

6.1.1 Vistula Lagoon is Hot Spot No. 73 in the HELCOM List

In June 2019, at the 56th meeting of the delegations of the HELCOM member states, hot spots No. 67 (Wastewater treatment plant in Kaliningrad) and No. 69 (Pulp and paper mill Zepruss) were excluded from the general list of hot spots (HELCOM, n.d.). Earlier, No. 68 (Darita pulp and paper mill) was excluded from this list. It contains seven more objects located in the Kaliningrad region, including two bays—Curonian and Vistula.

The Vistula Lagoon (Fig. 6.1) was included in the list of HELCOM hot spots due to the lack of effective resource management plans for these unique natural objects. Moreover, most of the Kaliningrad hot spots are directly related to the Vistula Lagoon. Some of them, as already mentioned, have officially ceased to be "hot", and some of them are close to losing this status.

However, tendencies in attitudes towards the natural systems of the Kaliningrad Region require not only a formal approach—whether they are included in the list or not, whether there are management plans for them or not. Due to sanctions and closure of many tourist destinations abroad, natural complexes of the Kaliningrad Region have become extremely popular in terms of tourism and recreation. The Vistula Lagoon is among these. Another important feature of the bay is its fish stocks popular with both professional and amateur fishermen.

In this case, we propose to consider the Vistula Lagoon as a water body exposed to long-turn chemical pollution from its catchment area posing a danger to both aquatic organisms and humans. In 2016, the Federal Law of the Russian Federation 7-FZ On

Environmental Protection introduced the concept of "accumulated ecological harm" and its definition: this is harm to the environment caused by past economic and other activities, where the obligation to eliminate the harm has not been met or has not been fully implemented.

The aim of this paper was to substantiate the inclusion of the Vistula Lagoon in the list of objects of accumulated ecological harm.

6.2 Materials and Methods

The paper employed articles and other available publications, which report data on the ecological status of the Vistula Lagoon and primarily information on the potential danger of chemicals accumulated in different environments of the bay to human.

6.3 Results

Municipal landfills are one of the main sources of water pollution. The largest landfill in the Kaliningrad Region was reclaimed in 2019. Two more municipal landfills are currently being prepared for reclamation. In 2005, there were 161 waste disposal sites in the Kaliningrad Region (Ministry of Natural Resources and Environment of the Kaliningrad Region 2018). By 2011, the number of landfills for disposal of household and industrial waste had reduced to 31 (Ministry of Natural Resources and Environment of the Kaliningrad Region 2012), and to date, the number of landfills used in the region has decreased to 4, of which only 2 are supplied with drainage water collection and purification systems.

The period of the reducing number of officially used landfills was accompanied by appearance of unauthorized landfills. For example, in 2011, 230 unauthorized landfills were discovered (Ministry of Natural Resources and Environment of the Kaliningrad Region 2012).

Another significant long-term source of water pollution in the Kaliningrad Region was household and industrial wastewater from settlements. The wastewater pollutants could enter the water bodies of the region.

At the beginning of the twentieth century, an intensive bloom of blue-green algae was periodically observed in the Vistula Lagoon, and both hydrobiologists and sanitary hygienists were aware of this fact. During these periods, fish deaths, diseases and deaths among the population consuming water from the bay were recorded. The disease was named Haff disease in honor of Frisches-Haff Bay (now Vistula Bay). At that time, the pollution of the estuary and urban parts of the Pregolya River attained the level threatening the life of aquatic organisms (Ezhova 2013).

Intensive algal bloom in the bay was also facilitated by the supply of nutrients from its catchment area involved in active agricultural activities. Since then, the list of leached nutrients has been extended by a variety of chemicals used in crop and livestock production, as well as pollutants from wastewater and filtration waters of the landfill.

Long-term regular monitoring of the influx of nutrients into the Vistula Lagoon and its aftereffects have been performed by the Atlantic branch of the Federal State Budgetary Scientific Institution VNIRO (AtlantNIRO), and its results are presented in numerous reports and reviews (Aleksandrov 2011, 2018; Aleksandrov et al. 2017). They state that the content of gross nitrogen and phosphorus over most of the years of the monitoring period, on average for the Russian water area, corresponded to the potentially hypertrophic level of the bay ecosystem.

These facts indicate intense eutrophication of the Vistula Lagoon. However, no systematic studies of other forms of chemical pollution in the bay have been performed until recently. Such studies have become more regular and their results have become available to the scientific community only in the last 15 years.

One of the main nutrient suppliers to the bay is the Pregolya River, as its basin makes up 65% of the catchment area of the Vistula Lagoon (Russian part of the bay) (Ejdel'man et al. 2020). The ecological status of the bay and the Pregolya was assessed based on the parameters of hydrochemical monitoring and meteorological reports obtained for 2013–2019. The actual content of chemical substances in the studied water objects was compared with the standards for the content of these substances in water bodies of fishery significance—maximum permissible concentrations (MPCr-x). During the entire monitoring period, annual excess of MPCr-x was recorded for ammonium ions, nitrite ions, chloride ions, iron, copper, sulfates, phenols, and formaldehyde.

The content of heavy metals in bottom sediments was of the greatest interest among researchers who investigated chemical pollution of the bay.

Arsenic contents in bottom sediments of the Vistula Lagoon have been determined (Emel'yanov and Kravcov 2007). In the 0–3 cm sedimentary layer, from 3 to 57 mg/kg of arsenic was found, and in the silt column of 5–66 cm, the arsenic content ranged from 9 to 133 mg/kg.

The mercury content in bottom sediments of some parts of the Vistula Lagoon was reported in the review prepared based on fragmentary data obtained for the period of 2000–2019 (Bogdanov et al. 2020b). Mercury content ranged from 0.008 to 0.216 mg/kg.

In individual cases (Pahomova et al. 2004), in the coastal Vistula Lagoon and in its central part, flows of dissolved inorganic metals (Fe II and III, Mn, Pb, Cu) and those associated with the organic matter of marine water (Fe, Mn, Pb, Cu) were studied at the water–bottom interface. Among other things, it was found that at the stations in the Primorskaya Bay located closer to the coast the metal fluxes are 2–fourfold higher than those at the station located closer to the bay center.

According to the Geology and Licensing Division of the Department for Subsoil Use of the Northwestern Federal District for the Kaliningrad Region (Efimenko and Sokolov 2015), over 350 water-consuming enterprises extract and use underground fresh and mineral waters for drinking, industrial and commercial purposes. The utility and drinking water system, which supplies the population of about 950 thousand people in the Kaliningrad Region with water, employs mainly 4 aquifers: upper intermoraine Moscow-Valdai, lower intermoraine Oka-Dnieper, Neogene-Paleogene and Upper Cretaceous. It is believed that the region is reliably provided with groundwater resources.

It is well known that surface waters can interact with groundwater with various degree of intensity. Extensive work has recently been done to assess this interaction in the Kaliningrad Region (Mihnevich 2011). As a result, three zones were identified in the region in terms of the degree of protection of the Moscow-Valdai aquifer (the most exploited) depending on the thickness of the upper aquiclude.

Hydrogeological (lithological) windows—local areas characterized by the absence or small thickness of the upper aquiclude, its increased permeability, and an insignificant water pressure in the Moscow-Valdai aquifer—provide interaction of different groundwater and surface water aquifers. Hydrogeological windows serve as pathways for pollutant spread in groundwater. To the greatest extent, hydrogeological windows are typical of the valleys of large rivers, for example, the Pregolya River.

The paper reports that there are 12 sanctioned (at the time of the study) landfills in the territory of the unprotected aquifers. Groundwater in 15 unauthorized landfills is not protected.

Another aftereffect of the influx of pollutants from land into groundwater aquifers is their subsequent migration (submarine discharge) to the Baltic Sea (Mihnevich et al. 2020). A special effect on the concentration of many heavy metals of submarine discharge is observed in small closed (semi-closed) water areas, namely gulfs and bays. In a number of such cases, the influx of metals such as cadmium, cobalt, zinc and chromium from the groundwater discharge can account for 80–90% of the river flow.

In (Bogdanov et al. 2020a), the ecological status of the Vistula Lagoon is assessed by the content of oil products in bottom sediments in three different zones of the bay. In conclusion the authors state that the ecological status of the investigated area of the Kaliningrad Bay is currently quite satisfactory in terms of the content of oil products in bottom sediments.

Studies are underway to investigate the aftereffects of chemical pollution of the aquatic environment by economic activities for aquatic organisms and human health.

It was found that in perch, roach and pike perch caught in the Curonian and Vistula Lagoons, the total PAH content does not exceed 11 μg/kg, while in bream, its content attains up to 14.6 μg/kg (Shenderyuk et al. 2011).

In general, all the most extensively studied pollutants of natural waters fall into three groups (Petrosyan 2010): priority organic toxicants, priority organometallic toxicants, and priority heavy metals.

In turn, priority organic toxicants include 13 groups of organic compounds with a wide range of chemical stresses: bioaccumulation in fatty tissues and in trophic chains, carcinogenesis, damage to the central nervous, endocrine systems, immune and reproductive systems, skin damage, toxic effects on fish etc.

Priority organometallic toxicants include derivatives of tin, mercury and lead. Chemical stresses of these compounds entail damage to the central nervous system, brain and liver, and cancer of the respiratory and digestive tracts.

Priority heavy metals in aquatic ecosystems include 12 metals with a wide range of chemical stresses.

Recent epidemiological developments draw our attention to the property of some metals that affects living organisms—mutagenesis, the ability to provoke new mutations. There are few studies on mutagenesis, therefore we can only state that metals in the aquatic environment can affect evolutionary processes at all levels of the organization of living matter, including viruses, bacteria, etc.

6.4 Discussion

Some of the promising economic activities in the Kaliningrad Region are fishing and fish processing, transport, and tourism. All these activities are somehow related to water, including water resources of the Vistula Lagoon.

In the Russian part of the Vistula Lagoon, more than 4 thousand tons of fish have been harvested in recent years, and the catch was represented by about ten fish species (herring prevail) (Shibaev 2020).

Water tourism, including sports tourism (sport fishing), is considered promising (Lekarkina 2017) due to the unique geographical position of the Kaliningrad Region. The Kaliningrad Region has a dense network of waterways, exits to the Baltic Sea and two bays. The development of cruise tourism and travelling by inland waterways is advantageous for the Kaliningrad Region.

The study of the Vistula Lagoon and associated waterways is highly potential for development of both water tourism and transportation of goods in the bay basin (Chubarenko et al. 2015). There are 19 small ports and harbors in the Polish and Russian waters of the bay.

A number of authors (Belova et al. 2016) believe that ecotourism is of a particularly high potential for development of tourist and recreational activities on the shores of the Vistula Lagoon, which combine efficiency of economic activities and high quality of the natural potential.

With regard to the uniqueness of most water bodies in the Kaliningrad Region, other authors suggest that plans for economic development of the region should include a management element referred to as ecological marketing of the territory (EMT, which indicates the priority of the environmental aspect in any business project implemented in the Kaliningrad Region (Kolosovskij 2017).

6.5 Conclusion

The Vistula Lagoon has been under long-term intense chemical exposure (at least for the last hundred years) caused by various chemicals entering it from the catchment area. Sources of chemical substances entering the aquatic environment are wastewater from settlements, industrial and agricultural enterprises and landfills. Chemicals enter the bay through surface runoff and groundwater (air transport is difficult to assess). In bottom sediments of the bay, high concentrations of various chemical compounds (primarily metals) pose a potential danger to aquatic organisms and humans. The bay is in high demand for economic activities, therefore, it is necessary to ensure its environmental safety, in particular, to reduce its chemical pollution and assess the potential danger to aquatic organisms and population of the region posed by chemical pollutants accumulated in the bay ecosystem.

The results of assessing the object of accumulated ecological harm should be sent to the Federal Government, which may include (or refuse) the object in the State Register (Decree of the Government of the Russian Federation of April 13, 2017 No. 445 On approval of the regulations for maintaining the State Register of objects of accumulated environmental harm). Objects included in the register are provided with state financial support for works aimed at elimination of the accumulated harm (Decree of the Government of the Russian Federation of May 4, 2018 No. 542 On approval of the regulations for organizing works to eliminate accumulated ecological harm), namely:

(1) adequate inspections of the facility, including engineering surveys;
(2) development of the project to eliminate the accumulated harm;
(3) agreement and approval of the project;
(4) works to eliminate the accumulated harm;
(5) control and acceptance of the work aimed at elimination of the accumulated harm.

References

Aleksandrov SV (2011) Current ecological state and pollution of the Curonian and Vistula bays of the Baltic Sea. Water: chemistry and ecology, 11, 3–9 (in Russian)

Aleksandrov SV (2018) Spatial changes in hydrochemical parameters in the Vistula Lagoon in 2014–2016. Works of AtlantNIRO, 2 1(5)), 5–21 (in Russian)

Aleksandrov SV, Vahrusheva SA, Mal'fanov IL, Trenina NE (2017) Spatial changes in hydrochemical parameters and water salinity in the Vistula Lagoon in 2010–2013. Works of AtlantNIRO. New series 1(3):5–21 (in Russian)

Belova AV, Fedina-ZHurbina IV (2016) Opportunities for the development of ecotourism in the Kaliningrad region (with an emphasis on the pilot area of the Russian part of the territory adjacent to the Vistula (Kaliningrad) and Curonian bays). Bulletin of the Baltic Federal University. I. Kant. Series: Natural and Medical Sciences 1:64–74 (in Russian)

Bogdanov NA, Bass OV, Savostina OA, and Chugaevich VYa (2020a) The ecological state of the kaliningrad bay by the variability of the content of oil products in bottom sediments. In V. I. Osipov

Proceedings from Sergeev's readings: geoecological aspects of implementation of the national project "Ecology". Dialogue of generations. Moscow, Russia: Peoples' Friendship University of Russia (in Russian)

Bogdanov NA, Ermolaev BV, Bass OV, Paranina AN (2020b) Mercury in bottom sediments of the water area and at river mouths: Kaliningrad Bay. Geosphere Research 2:76–82

Chubarenko BV, Shvankovska B, Fedorova GM, Gricenko VA (Eds.). (2015). Catalog of small ports and harbors of the Kaliningrad /Vistula Lagoon/. Moscow, Russia: Ex-libris. 188 p.

Efimenko AV, Sokolov AV (2015) Condition and rational use of groundwater resources in the Kaliningrad region. In K.V. Tylik (eds.) Proceedings from III BMF WBAEWB: Water bioresources, aquaculture and ecology of water bodies. Kaliningrad, Russia: Kaliningrad State Technical University (in Russian)

Ejdel'man YAV, Koroleva YuV, Golubeva YUV (2020) Comprehensive assessment of pollution of the Pregolya River and the water area of the Kaliningrad Bay. International Research Journal 6–2(96):74–78

Emel'yanov EM, Kravcov VA (2007) Reasons for increased arsenic concentrations in the Baltic Sea and Vistula Lagoon. Geochemistry 8:871–888 (in Russian)

Ezhova EE (ed.) (2013) Biological communities of the pregolya river (vistula gulf, baltic sea). Kaliningrad, Russia: Smartbooks. 246 p (in Russian)

HELCOM Hot Spots: two Kaliningrad pollution sites get wiped off the list at key HELCOM meeting. (n.d.). Retrieved from https://helcom.fi/helcom-hot-spots-two-kaliningrad-pollution-sites-get-wiped-off-the-list-at-key-helcom-meeting/

Kolosovskij AM (2017) On some aspects of environmental marketing of water resources in the Kaliningrad region. In Kostrikova N.A. (Eds.). Proceedings from V IBMF: V International Baltic Maritime Forum. Kaliningrad, Russia: Separate structural subdivision "Baltic State Academy of Fishing Fleet" of the Federal State Budgetary Educational Institution of Higher Professional Education "Kaliningrad State Technical University" (in Russian)

Lekarkina NK (2017) Overview of investment opportunities in the Kaliningrad region. Analysis of the most promising industries for investment. Economic scientific journal: Investment assessment, 2–1, (6), 55–80 (in Russian)

Mihnevich GS (2011) Geoecological assessment of the natural protection of groundwater from pollution (on the example of the system of the upper intermoraine aquifer of the Kaliningrad region). Dissertation of Candidate of Geographical Sciences. Immanuel Kant Baltic Federal University, Kaliningrad, Russia (in Russian)

Mihnevich GS, Krechik VA, Krek AV, Danchenkov AR (2020) Transport of heavy metals with underground runoff to the baltic sea. Proceedings from VIII IBMF: Materials of the VIII International Baltic Maritime Forum: in 6 volumes, 157–163 (in Russian)

Ministry of Natural Resources and Environment of the Kaliningrad Region (2018) Order of Natural Resources and Environment of the Kaliningrad Region "On Approval of the Territorial Scheme for Waste Management, Including Solid Municipal Waste, in the Kaliningrad Region" (March 28, 2018 No 145). Kaliningrad, Russia (in Russian)

Ministry of Natural Resources and Environment of the Kaliningrad Region (2012) Report "on the ecological situation in the Kaliningrad region in 2011". Kaliningrad, Russia. 203 p (in Russian)

Pahomova SV, MYu Kononec, YUdin MV, Vershinin AV, Rozanov AG (2004) Study of dissolved forms of metals across the water-bottom interface in the Vistula Lagoon of the Baltic Sea. Oceanology 44(4):516–523 (in Russian)

Petrosyan VS (2010) Natural and drinking water: chemical safety issues. RANS Bulletin 10(3):50–55 (in Russian)

Shenderyuk VV, Chernysheva NL, Baholdina LP, Dubova OL, Strygina OA (eds.) (2011) Assessment of the level of pollution of the water resources of the Baltic Sea and the waters of its sea and bays with polycyclic aromatic hydrocarbons. In AV Ivanov (eds.) proceedings from IX ISC : proceedings of the IX International Scientific Conference "Innovations in science and education—2011". Kaliningrad, Russia: Kaliningrad State Technical University (in Russian)

Shibaev SV (2020) Fishing intensity in the Russian part of the Vistula Lagoon. Fishing Issues 21(1):41–52 (in Russian)

Chapter 7
Invasive Species *Palaemon Elegans* Rathke, 1836, (Caridea: Palaemonidae) as the Only Species of Palemon Shrimps in Water Bodies of the Kaliningrad Region

S. A. Sudnik and Yu. E. Egorova

Abstract Globalization of the problem of the introduction of aquatic organisms apparently necessitates monitoring of the state of their fauna and primarily in open water bodies. For the first time, this study has assessed the species diversity of shrimp from the genus Palaemon in water bodies of the Kaliningrad region, which is relevant due to the ecological plasticity known for a number of species that promotes their successful dispersal and acclimatization. Palaemon elegans is known to have lived in this region since about 2000; Palaemon adspersus, Palaemon macrodactylus, and Palaemon varians, recorded in waters of Germany and Poland, are potential invaders into water bodies of the region, and their invasion has not been previously estimated. Species identification of 2498 individuals collected in the study water bodies in 2006–2018 revealed only one species—Palaemon elegans Rathke, 1836; the structure of the mandibles was an unambiguous distinguishing character for its identification; a number of other characters varied. The identified sporadic individuals (0.3%) with a two-segmented mandibular palp, which exhibited unusual combinations of other characters, were referred to as atypical representatives of P. elegans. Invasive P. elegans was currently the only palemon species in waters of the Kaliningrad region. An additional analysis of taxonomically important plastic and meristic characters of the structure of the rostrum, pereopods 2, and mandibles performed for 300 individuals of the species specified the diagnosis of Palaemon elegans Rathke, 1836, from water bodies of the Kaliningrad region. Now, it takes into account the morphological variation in shrimp from the southeastern Baltic Sea and the Kaliningrad Bay.

Keywords Invaders southeast · Baltic · Kaliningrad bay · Shrimp

---

S. A. Sudnik (✉) · Yu. E. Egorova
Kaliningrad State Technical University, Sovetsky prospect 1, 236022 Kaliningrad, Russian Federation
e-mail: svetlana.sudnik@klgtu.ru

Yu. E. Egorova
e-mail: yuliya.egorova@klgtu.ru

A. G. Arkhipov (ed.), *Sustainable Fisheries and Aquaculture: Challenges and Prospects for the Blue Bioeconomy*, Environmental Science and Engineering, https://doi.org/10.1007/978-3-031-08284-9_7

7.1 Introduction

Shrimp of the Palaemon genus are widespread in the World Ocean and belong to the family Palaemonidae, the leader among invasive species (Burukovsky 2017). Rockpool shrimp, Palaemon elegans Rathke, 1836, is an invasive species with a wide range due a high degree of eurybionticity (tolerant to salinity of 2–45 ‰, temperature of 2–25 °C), a wide food spectrum (benthophage, detritophage, necrophage), and planktonic larva. Having penetrated in the East Atlantic, the species invaded the Baltic, Mediterranean, Black, and Azov seas; it has been recorded in the Caspian and Aral Seas, the Suez Canal, the Red and Arabian Seas, and the Persian Gulf (Katajisto et al. 2013; CABI 2021). In 2010, P. elegans was found in the Western Atlantic (MIT Sea Grant College Program2010). Shrimp are found in the diet of fish, including commercial fish (European eelpout, round goby, flounder, cod) (Gruszka and Więcaszek 2004). It is an object of commercial and amateur fishing (European… 2021). Until the 1990s, the species was recorded in the western Baltic Sea (Köhn and Gosselck 1989), but in the early 21 century, probably due to human introduction, it widely spread in the southern and the northeastern Baltic Sea (Janas and Tutak 2014). Successful acclimatization of P. elegans in new water bodies and food competition may be accompanied by the displacement of native palemon species, as it happened in waters of Germany and Poland (Grabowski 2006). In 2000–2001, the species was first encountered in the Kaliningrad Bay and the southeastern Baltic Sea (Ezhova 2009), where it is now abundant. A number of studies report the aspects of P. elegans biology (Tsigvintsev 2008; Ezhova 2009; Burukovsky 2012; Gusev 2012; Miroshnichenko 2018; Sudnik and Egorova 2019). Three more species (Palaemon adspersus Rathke 1836, Palaemon macrodactylus MJ Rathbun 1902, and Palaemon varians (Leach 1814)) confined to waters of Germany and Poland (González-Ortegón 2006; Janas and Tutak 2014) could invade waters of the Kaliningrad region; however, no special studies of the palemon fauna in the region have been performed. Data on biodiversity and distribution of hydrobionts are of huge significance in developing monitoring programs for alien species to manage water systems. We assessed taxonomically significant plastic and meristic characters for 2498 shrimps from the southeastern Baltic Sea and the Kaliningrad Bay and found an almost absolute predominance of Palaemon elegans Rathke, 1836. Individuals with a two-segmented mandibular palp (0.3%) that exhibited unusual combinations of other characters were referred to as atypical representatives of P. elegans; the mandible structure was an unambiguous character for species identification. Invasive P. elegans is currently the only palemon species in waters of the Kaliningrad region.

7.2 Materials and Methods

The aim of this study was to analyze the species diversity of shrimp Palaemon spp. in waters of the Kaliningrad region. Material: 2498 palemon individuals collected by the authors and employees from Kaliningrad State Technical University in 2006–2018 using a shrimp dredge (1 m wide, 0.5 m vertical opening, 5 mm mesh) in the southeastern Baltic Sea, and with an ISO 10870:2012 hydrobiological landing net (0.5 mm mesh) (Table 7.1) in the Kaliningrad Bay; fixation with 8% formaldehyde solution. For species identification, a key was used (Burukovsky 2017) with the diagnoses of the species taken from data by (Köhn and Gosselck 1989), who described them for shrimp from water bodies in Germany.

Species identification of 2498 individuals (Table 7.2: general analysis) was performed with regard to the species characters: the rostrum size, the distribution of teeth on its sides, the structure of the mandible palp, cheliferous legs (pereopods 2). For identification of species Palaemon elegans, the following diagnosis was used: the rostrum is straight or slightly curved upward; the maximum width of its ventral surface is greater than that of the dorsal one; the rostrum length is approximately equal to the carapace length, its distal end does not extend beyond the distal edge of scaphocerites, it is bifurcated; the rostrum dorsal side with 7–10 (usually 8–9) teeth, of which 3 (rarely 2) are located on the carapace behind the eye orbit; the underside of the rostrum with 3–4 teeth; mandibles with a two-segmented palp; in pereopods 2, carpus are longer than merus and shorter than propodus; the dactylus length is 0.4 of the manus length.

For 300 individuals identified as Palaemon elegans, additional analysis was performed (Table 7.2) to specify the diagnosis of the species from water bodies of the Kaliningrad region. It included assessment of a number of plastic and meristic characteristics of the structure of the rostrum and pereopods 2. An MBS-10 binocular micrometer with an accuracy of 0.01 mm was used to measure the length of the exoskeleton segments (see Fig. 7.1): carapace (CL; from the lower edge of the eye

Table 7.1 Palemon shrimp material

Collection period, year	Water body	Depth, m	Number of individuals, ind.	
			General analysis	Additional analysis
2006—2018	Kaliningrad Bay, vistula spit, kosa VILL	0.1–0.5	2171	237
2009	Baltic sea, Kulikovo VILL	1.0–1.5	81	
2010, 2014, 2015	Baltic sea, Rybachy VILL, Donskoye VILL	0.1–1.0	246	63
Total			2498	300

Source Compiled by the authors

Table 7.2 Results of species identification of palemon shrimp from water bodies in the Kaliningrad region

Collection period, year	Water body	Number of individuals, ind. (% in harvest)	
		Typical representatives of *Palaemon elegans*	Atypical representatives of *Palaemon elegans*
2006–2013, 2018	Kaliningrad bay	1302	–
2015, 2016		864	5 (0.9%)
2009	Baltic sea, Kulikovo settlement	78	3 (3.8%)
2010, 2014, 2015	Baltic sea, Rybachy settlement, Donskoye settlement	246	–
Total		2490 (99.7%)	8 (0.3%)

Source: Compiled by the authors

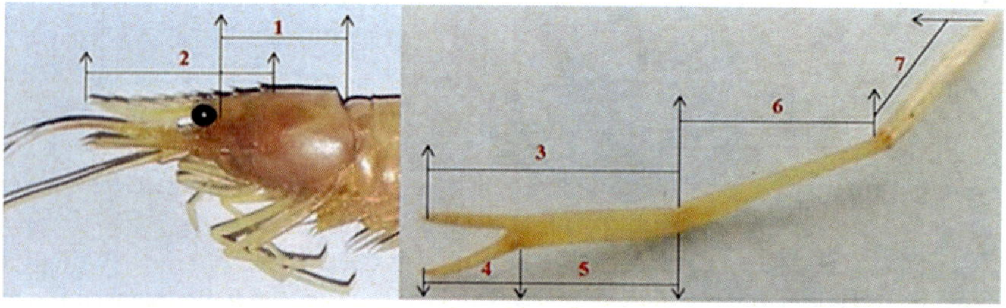

Fig. 7.1 Measuring the length of shrimp exoskeleton segments (orig.): 1—carapace; 2—rostrum; pereopods 2: 3—propodus; 4—dactylus; 5—manus; 6—carpus; 7—merus (orig.)

orbit to the middle of the posterior carapace edge), rostrum (LR; from the end of the rostrum to the eye orbit; the procedure for measuring the length of pereopods 2 segments (propodus (Lp), manus (Lm), dactylus (Ld), carpus (Lc), merus (Lmer)) is shown in Fig. 7.1. The analysis of the rostrum structure assessed the variability in its size (variations in the LR / CL index), data on the number of teeth on the rostrum and their distribution on its dorsal and ventral sides, which was expressed by formula (1) compiled by the authors based on the description of the rostrum structure in the diagnosis of P. elegans:

$$x + y + z / v \tag{7.1}$$

where x is the number of subapical teeth; y, z are the number of dorsal teeth located before (y) and behind (z) the posterior edge of the eye orbit; v is the number of ventral teeth (see Fig. 7.2).

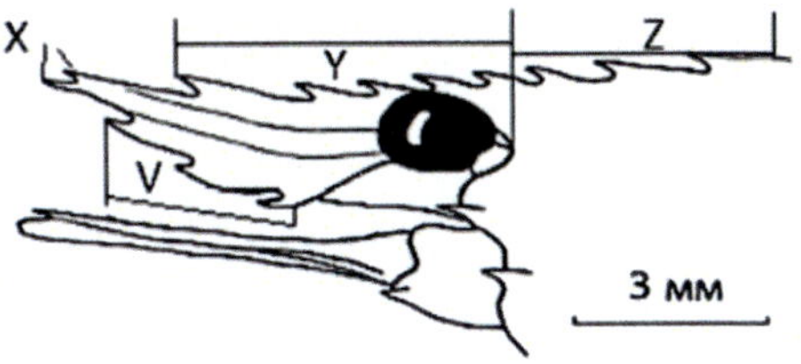

Fig. 7.2 Designations of the teeth in the formula for the rostrum of palemon shrimp (orig.)

The study of the structural features of pereopods 2 assessed the variability in the size of the carpus (relative to the sizes of the propodus and merus using the indices: Lc / Lp and Lc / Lmer) and dactylus (relative to the size of the manus using the index Ld / Lm), in accordance with the diagnosis of the species.

The variation in characters was evaluated by standard statistical methods "Microsoft Office Excel".

7.3 Results

The species identification performed for 2498 palemon individuals from water bodies in the Kaliningrad region revealed virtually absolute predominance of *Palaemon elegans* Rathke, 1837, with characters typical of this species (Table 7.2).

In addition, sporadic individuals (Table 7.2) found in the Baltic Sea in 2009 and at the entrance to the Kaliningrad Bay in 2015–2016 corresponded to the genus Palaemon (Burukovsky 2017) in appearance, but differed in characters from Palaemon elegans and other palemon species, supposed invaders in water bodies of the region (P. adspersus, P. macrodactylus, P. varians). The rostrum of the shrimp was different in shape and size (the LR/CL index could be less than 0.7), in number (only 6–8 or 12 dorsal teeth), in the pattern of teeth distribution on the rostrum, and in the structure of the claw of pereopods 2 (the length of the dactylus attained 0.7 and more of that of the manus); one individual could exhibit a combination of characters typical of different species. This indicates morphological variation of the studied characters. However, all the individuals had mandibles with two-segmented palps, which refers them to atypical representatives of the species Palaemon elegans.

Description of typical representatives of Palaemon elegans from water bodies in the southeastern Baltic Sea and the Kaliningrad Bay

All the individuals had mandibles with two-segmented palps. Tables 7.3 and 7.4 summarize the results of assessing the variability in the rostrum size (variation in the LR/CL index), and the number and position of teeth on the rostrum (according to its formula). In most of the individuals, the rostrum length was slightly less than that of the carapace; however, the differences found were insignificant (Table 7.3), which correlates with the data obtained for species from Germany (Köhn and Gosselck 1989): "the length of the rostrum and the carapace are approximately equal". The

Table 7.3 Characteristics of the rostrum size in shrimp *Palaemon elegans*

Variability in the relative size of the rostrum	Share of individuals, %	Index LR / CL	
		Range	Mean ± standard deviation
Rostrum length = carapace length	0.3	1.00	–
Rostrum length < carapace length	85.2	0.34–0.99	0.82 ± 0.14
Rostrum length > carapace length	14.5	1.01–1.34	1.06 ± 0.14

Source Compiled by the authors

Table 7.4 Characteristics of the rostrum formula in shrimp *Palaemon elegans*

Rostrum formula	2 + 4 + 3/3	2 + 5 + 2/ 3	2 + 5 + 3/3	2 + 6 + 2/3	2 + 6 + 3/3	2 + 5 + 3/4	2 + 6 + 3/4
Share of individuals, %	47.3	17.9	28.9	3.8	1.2	0.6	0.3

Source Compiled by the authors

number of dorsal teeth (including 2 permanent subapical teeth) attained 9 in 65.2% of the total number of individuals, 10 in 33.3% of the individuals, and 11 in 1.5% of the individuals. The number of teeth located before the posterior edge of the orbit amounted to 4 and 5 in 47% of the individuals, and 6 in 5% of the individuals. The number of teeth behind the posterior edge of the orbit amounted to 3 in the majority (78%) of shrimps, and 2 in the rest (22%) of the individuals. Almost all the individuals (99%) had 3 ventral teeth, and some of them had 4 ventral teeth (Table 7.4).

Table 7.5 presents the results of the analysis of taxonomically significant indices in P. elegans, which show the segment length ratios for pereopods 2. In 85–89% of the individuals, the carpus was longer than the merus and shorter than the propodus (or equal to it); the individuals with the dactylus length less than the manus length dominated (97%). In general, these data were similar to those obtained for P. elegans from water bodies in Germany (Köhn and Gosselck 1989): "the carpus of pereopods 2 is typically longer than the merus and shorter than the propodus; the dactylus is much shorter than the manus."

The structure of the rostrum and pereopods 2 in P. elegans was found to vary, and a number of its specific characters were similar to those of other 3 palemon species that can inhabit this region, therefore, the structure of their mandibles should be assessed to identify palemon species from water bodies of the Baltic Sea basin: a two-segmented palp is an apparent and sufficient character of P. elegans species; the absence of the palp is characteristic of P. varians; in case of a three-segmented palp of mandibles, the remaining characters should be assessed to identify the species: P. adspersus or P. macrodactylus (Burukovsky 2017; Van Couwelaar 2003). Our data

Table 7.5 Characteristics of the structure of pereopods 2 in *Palaemon elegans*

Length ratio of segments for pereopods 2	Share of individuals, %	Index value	
		Range	Mean ± standard deviation
		Lc / Lmer	
Carpus length > merus length	85	1.02–1.23	1.15 ± 0.04
Carpus length < merus length	15	0.87–0.98	0.94 ± 0.08
		Lc / Lp	
Carpus length ≤ propodus length	89	0.41–1.00	0.75 ± 0.19
Carpus length > propodus length	11	1.03–1.50	1.15 ± 0.19
		Ld / Lm	
Dactilus length ≤ manus length	97	0.18–1.00	0.50 ± 0.22
Dactilus length > manus length	3	1.06–1.55	1.26 ± 0.24

Source Compiled by the authors

specify the diagnosis of species Palaemon elegans Rathke 183 from water bodies in the Kaliningrad region: mandibles with a two-segmented palp; rostrum length is 0.86–1.06 of the carapace length, its distal end does not extend the distal edge of scaphocerites, 2 subapical teeth, 9–11 (typically 9–10) teeth on the dorsal side, of which 3 (rarely 2) teeth are located behind the eye orbit, 3 (rarely 4) teeth on the ventral side. In pereopods 2, the carpus length is 1.15 ± 0.04 of the merus length and 0.75 ± 0.19 of the propodus length; dactylus length attains 0.50 ± 0.22 of the manus length.

7.4 Discussion

Globalization of the problem of species introduction and extensive transportation communication between the countries of the Baltic region necessitate regular monitoring of the fauna of palemon shrimp in water bodies of the southeastern Baltic. At present, the invasive species Palaemon elegans Rathke, 1836, is the only species from the genus Palaemon in water bodies of the Kaliningrad region. The prerequisites for dispersal of new species (primarily P. adspersus, P. macrodactylus, and P. varians) in water bodies of the region include their high tolerance to a wide range of temperature and salinity, hypoxic conditions (Grabowski 2006; Janas and Tutak 2014), and parental care behavior, which enhance their survival, dispersal and acclimatization capacity. Dispersal of these species in water bodies of the Kaliningrad region can be restrained by narrower limits of the tolerance to environmental conditions in their larvae, the confinement of adults to certain soils. We have elaborated the basis for a biometric passport for the species Palaemon elegans from the southeastern part of the Baltic Sea and the Kaliningrad Bay; the above-mentioned interspecific differences will facilitate and simplify the identification of shrimp species from the genus Palaemon, and will help assess the invasion pathways of species and analyze populations. Further studies into various aspects of the rockpool shrimp biology are of relevance, since it is sufficiently eurybiontic to the whole range of conditions, significant for the detrital food chain and for the diet of a number of commercial fish, and an object of recreational fishing.

7.5 Conclusion

The invasion of new species into the established hydrocenoses is an urgent problem in theoretical and applied hydrobiology. The species identification of shrimp from the genus Palaemon collected in the coastal waters of the southeastern Baltic Sea and the Kaliningrad Bay in 2006–2018 confirmed the presence of only Palaemon elegans Rathke, 1836, out of four palemon species that can be found in these waters. The species inhabited these waters in the late 1990s, and now it is widespread along the coast of the region. Its successful dispersal and acclimatization in the Baltic Sea are facilitated by its high tolerance to salinity and temperature, a wide food spectrum with the dominance of plant detritus that is abundant in coastal waters, and the ability of its larvae to withstand low salinity of the desalinated waters in the Baltic Sea and the Kaliningrad Bay. This helps the species to occupy new territories, to displace and replace other species in competition for food and in unstable shallow-water conditions of the littoral zone, and makes it an interesting model object for complex studies of invasive and ecological processes in hydrobiocenoses and a candidate for coastal aquaculture.

References

Burukovsky RN (2012) On the diet of shrimp Palaemon elegans Rathke 1837 (decapoda, palaemonidae) in the vistula lagoon. Journal of the Siberian Federal University 2:151–159

Burukovsky RN (2017) The shrimp genus Palaemon (crustacea, decapoda, palaemonidae) of the European seas: species identification. Kaliningrad: BGA RF FGBOU VO "KSTU"

CABI (2021) Invasive species compendium. Wallingford, UK: CAB International. Retrieved from https://www.cabi.org/isc

European Commission website (2021) Retrieved from https://fish-commercial-names.ec.europa.eu/fish-names/species_en?sn=26413

Ezhova E (2009) Abstract book of the 7th Baltic Sea Science Congress. Tallinn University of Technology

González-Ortegón E, Cuesta JA, Schubart CD (2006) First report of the oriental shrimp palaemon macrodactylus Rathbun, 1902 (Decapoda, Caridea, Palaemonidae) from German waters. Helgol Mar Res 61:67–69

Grabowski M (2006) Rapid colonization of the Polish baltic coast by an Atlantic palaemonid shrimp Palaemon elegans Rathke, 1837. Aquat Invasions 1:116–123

Gruszka P, Więcaszek B (2004) Book of abstracts, Baltic—the Sea of Aliens. Gdynia, Poland

Gusev AA, Rudinskaya LV (2017) Zoobenthos fauna of the South-Eastern Baltic Sea (Gdansk Basin) at different research periods. Trudy AtlantNIRO 3:33–64

Janas U, Tutak B (2014) First record of the oriental shrimp Palaemon macrodactylus MJ Rathbun, 1902 in the Baltic Sea. Oceanol Hydrobiol Stud 4:431–435

Katajisto T, Kotta J, Lehtiniemi M, Malavin SA, Panov VE (2013) Palaemon elegans Rathke, 1837 (Caridea: Palaemonoidea: Palaemonidae) established in the Gulf of Finland. BioInvasions Records 2:125–132

Köhn J, Gosselck F (1989) Bestimmungschlüssel der Malacostraken der Ostsee. Mitt Zool Mus Berl 65:3–144

MIT Sea Grant College Program (2010) State Researchers Find New Non-Native Invasive Species. Retrieved from http://seagrant.mit.edu/news/press_releases.php?ID=150

Miroshnichenko YaV (ed.) (2018) Materials of interuniversity scientific-technical conference of students and cadets '18: "Days of Science". BGA RF FGBOU VO "KSTU", Kaliningrad, Russia, 160–165

Sudnik SA, Egorova YuE (2019) Materials of the international scientific conference '19: Baltic maritime forum. FGBOU VO "KSTU", Russia Kaliningrad, 291–300

Tsigvintsev SV (2008) Problems of ichthyopathology and hydrobiology: the first steps in science. FGOU VPO "KSTU", Kaliningrad, Russia, 180–201

Van Couwelaar M (2003) Zooplankton and Micronekton of the North Sea. Marine species Identification Portal. Retrieved from http://speciesidentification.org/species.php?species_group=zmns&id=492

Chapter 8
Influence of Water Quality in Urban Water Reservoirs of Kaliningrad on the State of Amateur Fishing

N. A. Tsupikova, E. A. Sevostianova, and A. S. Popova

Abstract This paper is based on the data of hydrological and hydrochemical monitoring of urban water bodies in Kaliningrad and survey of amateur fishermen. It presents an analysis of influence of the ecological state of popular artificial ponds (the Verkhny, the Nizhny, the Poplavok, and the Forelevy) on the amateur fishing. Amateur fishing with recreational purposes is one of the most popular hobbies of the population in Kaliningrad, attracting representatives of various age and social groups. Such use of recreational facilities, on the one hand, enhances public interest in environmental problems and, at the same time, significantly depends on the state of resources. The reservoirs considered in the article belong to the meso-eutrophic type, which might degrade their attractiveness for amateur fishermen due to the low abundance of fish species. The results obtained made it possible to determine the main fish species caught in the studied water bodies, including typical inhabitants of eutrophic slow water flow or stagnant pools (roach, perch, crucian carp, bream, silver bream, ruff, pike). The paper considers both the average annual and seasonal distribution of amateur anglers across water bodies, the rating of water bodies for fishing and its explanation. The main criteria for attracting the population to recreational fishing are highlighted. It was found that amateur anglers prefer fishing spots that are distant from the crowded public areas. The most popular are large reservoirs, allowing people to comfortably distribute along the lakeshore for recreation.

Keywords Urban recreational fishing · Distribution of amateur anglers · Ecological state of water bodies

N. A. Tsupikova (✉) · E. A. Sevostianova · A. S. Popova
Kaliningrad State Technical University, Sovetsky prospect 1, 236022 Kaliningrad, Russian Federation
e-mail: tsoupikova@klgtu.ru

E. A. Sevostianova
e-mail: ekaterina.lozitskaya@klgtu.ru

A. S. Popova
e-mail: aleksandra.popova@klgtu.ru

A. G. Arkhipov (ed.), *Sustainable Fisheries and Aquaculture: Challenges and Prospects for the Blue Bioeconomy*, Environmental Science and Engineering, https://doi.org/10.1007/978-3-031-08284-9_8

8.1 Introduction

Amateur fishing (in foreign literature, usually "recreational") is one of the most popular hobbies of the population for outdoor recreation. Under Russian legislation, recreational fishing is understood as the activity for the extraction (catch) of aquatic biological resources carried out by citizens to meet their personal needs, as well as during official leisure and sports events (Federal Law 2018). The lack of global monitoring and statistics collection on recreational fishing activities and catches limits the possibility to understand the size and needs of this fishing sector. According to FAO data for 2012, about 7% of the population in developed countries is adherents of such leisure (FAO 2018). This suggests that in Kaliningrad, whose population is about half a million, at least 30 thousand people are engaged in recreational fishing. There has even been a recent trend in inland waters towards a gradual replacement of commercial fishing by recreational one (FAO, 2010). Moreover, recreational fishing in many cases can provide a much more economically efficient use of aquatic biological resources in comparison with commercial fishing (Shibaev and Sokolov 2014).

Recreational fishing is an important element of human interaction with nature, and this leisure activity is most demanding on natural water quality. In addition, fishermen note that in recreational fishing they value the opportunity to relax, spend time outdoors, enjoy sports, catch fresh fish for food, and spend time with friends (McInnes et al. 2013). This allows recreational fishing to be considered a part of the experience economy, an important component of the currently booming blue bioeconomy concept.

Such a pursuit helps maintain health in today's society. Recreational fishing does not imply congestions; fishermen, on the contrary, value privacy, which is important in a pandemic. Small urban water bodies surrounded with green spaces, which are located within easy walking and transport accessibility, are of particular value, especially during periods of lockdown.

One of the main factors limiting the recreational use of water areas is water pollution (Vasiliev and Kukushkin 1988). The fishermen consider the reservoirs with at least several fish species to be more attractive fishing grounds. Failure of water bodies to comply with fishery requirements, changes in hydrological conditions lead causes a decrease in the species diversity and number of fish, disruption of the reproduction of fish stock, as well as deterioration in the nutritional quality of fish meat. The presence of fish in inland waters is one of the best indicators of water quality.

Most of Kaliningrad's water bodies are small, often closed and stagnant, and experience significant anthropogenic pressure. Increasingly, the media publishes information about their pollution, which entails the death of fishes and birds. This significantly reduces the recreational attractiveness of water bodies for the population. So, only during 2021, mass fish mortality was observed in the Verkhny pond, the Nizhny pond and the Forelevy pond.

Purpose of this work is to determine the importance of recreational fishing as one of the possible criteria for assessing the ecological state of urban water bodies.

8.2 Materials and Methods

This paper analyzes the impact of the ecological state of water bodies on amateur fishing in four artificial reservoirs in Kaliningrad. The objects of the study are the Verkhny pond, the Nizhny pond, the Poplavok pond and the Forelevy pond (see Fig. 8.1).

(1—the Verkhny pond, 2—the Nizhny pond, 3—the Poplavok pond, 4—the Forelevy pond).

The water quality of the ponds was assessed according to the data of hydrological and hydrochemical monitoring, which has been connducted out by the authors since 2015. Monthly sampling from the surface is carried out at each of the studied water reservoirs. Water samples were analyzed in the KSTU hydrochemical laboratory using standard methods (Ryabtseva and Tsupikova 2019). Water trophic state and saprobity of the ponds was determined in compliance with the taxation rules established by the state standard (State Standard, 1978).

The registration of amateur fishermen with the calculation of the volume of their fish catch was performed at the water bodies of Kaliningrad during the period from

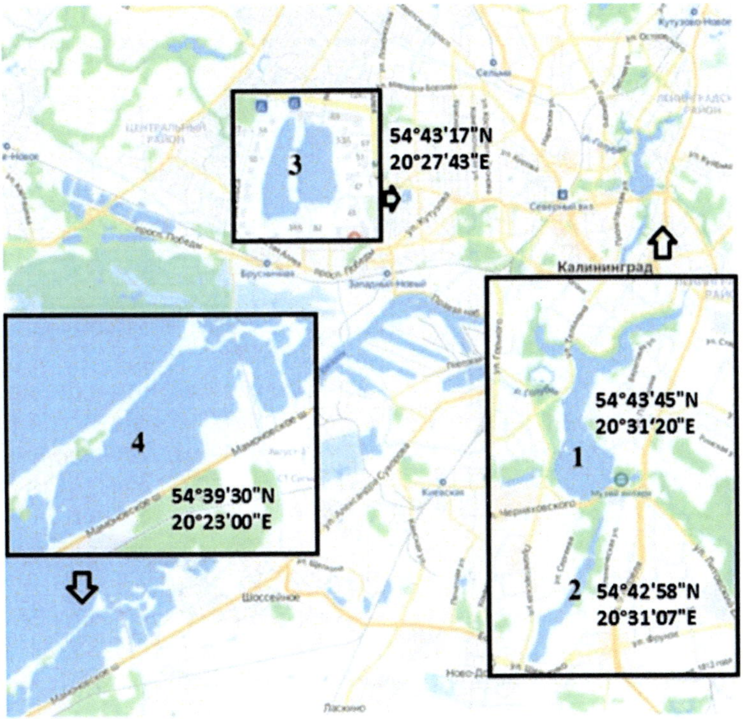

Fig. 8.1 Location of the studied reservoirs (yandex maps)

September 2020 to September 2021. Data collection was carried out by visual observation with photographs and in the form of a questionnaire-based survey of amateur fishermen, which allowed to assess the ponds attendance, intensity of fishing, duration of fishing and species composition of catches. Standard techniques were applied for mass measurements (Pravdin 1966).

8.3 Results

Hydrological and hydrochemical monitoring and registration of amateur fishermen revealed the results shown in the Tables 8.1–8.3.

On average, within one inspection of amateur anglers at the Forelevy pond, there were found from 2 to 10 fish specimens in the catches of each fisherman, at the Nizhny pond, there were slightly less—from 2 to 6, at the Verkhny pond—1–4 specimens, and at the Poplavok pond—from 3 to 11. The catches of amateur fishermen are dominated by roach (53.4% of the total number in catches), crucian carp (23.2%) and perch (11,6%).

Table 8.1 Hydrological parameters in studied water bodies, average for growing season

Parameter		Pond			
		Forelevy (2018)	Verkhny (2019)	Nizhny (2019)	Poplavok (2019)
Dissolved oxygen	mg/l	8,8 6,7–11,0	9,3 7,8–10,8	8,9 5,5–12,9	6,8 5,1–8,4
	%	100 77–124	104 90–123	106 64–151	73 56–85
Permanganate value	mgO/l	13,1 11,4–14,4	12,7 7,9–16,6	16,9 13,4–20,9	6,4 3,1–9,0
Phosphate	mgP/l	0,01 0,01–0,02	0,05 0,04–0,08	0,07 0,04–0,10	0,00 0,00
Iron (incl. ferrous and ferric iron)	mgFe/l	0,07 0,03–0,11	0,23 0,15–0,33	0,37 0,26–0,76	0,11 0,0–0,48
Ammonia nitrogen	mgN/l	0,58 0,31–0,93	0,63 0,18–1,17	0,30 0,21–0,41	0,16 0,0–0,41
Nitrite	mg/l	0,01 0,0–0,02	0,04 0,01–0,09	0,05 0,01–0,12	0,01 0,0–0,01
Water temperature	°C	22,5 21,6–23,0	20,3 18,0–23,5	20,2 12,4–23,9	19,5 13,4–22,1

Note Above the line is the average value, below the line is the fluctuations range

Table 8.2 Average annual results of registration of amateur fishermen

Parameter		Pond			
		Forelevy	Verkhny	Nizhny	Poplavok
Average (for the period) amount of amateur fishermen attending a pond, persons	Entire year	7,0	4,1	4,0	1,6
	Summer	12,4	4,5	6	2,5
	Autumn	6,8	2,8	5,3	1,2
	Winter	1,0	5,5	0,0	0,0
	Spring	5,9	2,3	4,3	2,0
Average (for the period) density of amateur fishermen per 1 km of coastline, persons / km	Entire year	0,88	0,67	1,67	1,76
	Summer	2,5	0,74	2,5	1,04
	Autumn	0,84	0,46	1,2	0,52
	Winter	0,13	0,90	0,0	0,00
	Spring	0,74	0,38	2,3	0,83

Table 8.3 Fish species in the catches of amateur fishermen in 2020–2021

Pond	Fish species, specimen/per cent						Total, specimen
	White bream	Crucian carp	Pike	Roach	Perch	Bream	
Forelevy	47 / 13,0	–	5 / 1,4	237 / 65,7	50 / 13,8	22 / 6,1	361
Verkhny	–	1 / 1,1	–	68 / 73,1	24 / 25,8	–	93
Nizhny	–	78 / 69,0	–	34 / 30,1	1 / 0,9	–	113
Poplavok	–	71 / 87,6	2 / 2,5	8 / 9,9	–	–	81
Total	47 / 7,3	150 / 23,2	7 / 1,1	347 / 53,4	75 / 11,6	22 / 3,4	648

8.4 Discussions

In numerous reservoirs of Kaliningrad, mostly eurybiotic fish species are widespread, such as roach, perch, crucian carp, bream, silver bream, ruff. The common carp (Carassius carassius L.), roach (Rutilus rutilus L.), perch (Perca fluviatilis L.), and pike (Esox lusius L.) become the objects of amateur fishing more often than others.

Crucian carp prefers silted stagnant water bodies and can withstand extreme environmental conditions, remaining the only representative of ichthyofauna in reservoirs suffering fish kills. Roach and perch, omnipresent in the water bodies of the Kaliningrad region and often found in the catches of fishermen at the studied urban ponds, form perch-roach reservoirs. The most valuable fish species for amateurs is pike. However, in the studied ponds, mainly small specimens are caught, which cannot be used further.

All investigated reservoirs are of artificial origin. The Verkhny and the Nizhny ponds are among the oldest ones in the region (both built in the thirteenth century). The Verkhny pond was used for fish farming, the Nizhny pond was first intended to fill the moats of the Konigsberg castle with water, and later it became a source of water supply for the city (Gubin 1997).

The Poplavok and the Forelevy ponds are much younger. The Poplavok pond was created at the beginning of the twentieth century in order to avoid flooding of the adjacent territories. The Forelevy pond is a part of the Golubye Lakes system, emerged in the 1970s in the former sand and gravel quarries. Since 1976, a trout farm has been operating on the north-western coast of the reservoir.

The Verkhny, the Nizhny and the Poplavok ponds are located in the city center, they have been reconstructed, so these places are popular both for walking and for comfortable amateur fishing, while the Forelevy pond is situated on the southwestern periphery of Kaliningrad. Compared to the reservoirs in the city center, it is less comfortable though due to the rather long coastline (6 km) and a trout farm, it is especially interesting to amateur fishermen. Currently, the studied water bodies mainly perform recreational and aesthetic functions.

The reservoirs differ in morphological parameters. Small ponds the Poplavok (water surface area 2.3 hectares) and the Nizhny (8.5 hectares) are shallower, their depths do not exceed 3.5 m. The larger Verkhny pond (29 hectares) and the Forelevy pond (about 90 hectares) are significantly deeper, for instance, the depth of the Forelevy pond is more than 20 m.

According to the analyzed hydrochemical parameters, the waters of the Poplavok and the Forelevy ponds are cleaner, oligosaprobic and oligo-betamezosaprobic, while the Verkhny and the Nizhny ponds are polluted, betamezosaprobic. Almost all the studied water bodies are eutrophic; water quality in the Poplavok allows to classify it as mesotrophic, which is probably due to recent cleaning and reclamation (Tsupikova and Sevostianova 2021).

The content of dissolved oxygen during the growing season is quite high (on average 8.8–9.3 mg/l) in all the ponds under consideration, except for the Poplavok, in which the average oxygen concentration is 6.8 mg/l. It is noticeably lower than in other reservoirs. Permanganate value changes in ponds in a similar way. It is the smallest in the Poplavok (moderate oxygen requirement), but it increases in the Verkhny, the Nizhny and the Forelevy ponds (12.7–16.9 mgO/l).

High concentrations of ammonium nitrogen are unfavorable for ichthyofauna as both its ion and the reduced form (ammonia) are capable of reacting with proteins and causing their denaturation. The excess of permissible concentrations of ammonium nitrogen was observed in all water bodies: occasionally in the Poplavok and in the Nizhny ponds, in the Verkhny and the Forelevoe ponds—many times throughout the observation period (up to 3–4 MPC). This is especially dangerous at high water temperatures, which did not fall in the water bodies under consideration during the growing season below 20 °C. Penetrating through the integuments and the branchial apparatus into the blood, nitrites form methemoglobin causing hemic and histotoxic hypoxia accompanied in severe cases by the death of fish and other reservoirs inhabitants (Shakhnazarova 2005). The content of nitrites exceeding the MPC was found in the Verkhny-Nizhny pond system.

According to the research carried out, the Forelevy pond is the most attended by fishermen on average for a year—up to 16 persons for the entire reservoir within one inspection. However, in terms of the average density of amateur fishermen per 1 km of the coastline, the small ponds Poplavok and Nizhny take leading position (1.6–1.7 persons/km). Seasonal distribution shows that the largest number of fishermen could be found in summer (more than 2.5 persons/km at the Forelevy). In winter, amateur fishermen preferred the Verkhny pond (0.9 persons/km in winter with 0.7 persons/km on average for a year) and could not be met at all at the shallow Nizhny and the Poplavok.

The popularity of the Forelevy pond can be explained by the diversity of its ichthyofauna and the presence of a trout farm. Fishermen are attracted by the possibility of catching fish species uncommon for Kaliningrad (trout, sterlet) bred in the farm. The configuration of the farm's cages does not imply the release of juveniles into the reservoir, but this cannot weaken the feeling of fishing excitement among amateurs and is an additional factor in the attractiveness of the reservoir for recreation and fishing. It is inhabited by such species as roach (caught in the largest quantities), perch, silver bream, bream and pike, sometimes eel, trout. Since there are no approaches to the water on the reservoir, trampled places and new paths have appeared along its shore over time, soil and vegetation have been damaged by vehicles.

The next most visited pond is the Nizhny, with a landscaped embankment and convenient approaches to the water, which makes it easy to find a place for fishing. The peak of attendance of the reservoir is observed in summer when the density of amateur fishermen reaches 2.5 persons/km. The catches of amateur fishermen include roach, perch, crucian carp (the most common species) and sometimes tench.

There are pedestrian and bicycle paths along the perimeter of the Verkhny pond, popular for winter fishing, and there are convenient approaches to the water everywhere. But the attractiveness of the reservoir for the population and the liveliness on the embankments reduces the number of amateur fishermen visiting it in the warm season. Fishing is carried out in less passable places. The catches of amateurs included roach (in the largest quantities) and perch.

The Poplavok is no less comfortable and accessible for fishing than the Verkhniy and the Nizhniy ponds. Due to the smallest area, its attendance rates are low, on average per year—one or two fishermen for the entire reservoir, and in winter the pond is not used. In addition, the Poplavok is characterized by a very poor ichthyofauna, related, among other things, to the recent work on the reconstruction of the reservoir including its drainage. In the catches of amateur fishermen, mainly crucian carp (in the largest quantities) and small-sized roach are present. No large fish specimens were found in the reservoir.

8.5 Conclusion

(1) Survey of amateur anglers showed that the popularity of the reservoir for recreation of the population reduces its attractiveness for fishermen. Urban reservoirs with increased traffic are preferred for short-term leisure by people with limited mobility (seniors, families with small children, teenagers). In general,

recreational anglers are more likely to choose larger water bodies that allow optimal and comfortable distribution along its coastline.

(2) According to the research results, on average for a year, the largest Forelevy pond is the most attended, up to 16 persons for the whole reservoir. The highest average annual density of amateur fishermen per 1 km of coastline was observed at the small ponds Poplavok and Nizhny (1.6–1.7 persons/km). The Verkhny and the Forelevy ponds, due to their larger size, have slightly lower indices (0.7–1.0 persons/km). Uneven seasonal distribution of amateur fishermen is characteristic for attendance of water bodies. Only the Verkhny pond is used in winter, due to its accessibility and sufficient depth; in this season, the density of fishermen here is 0.9 persons/km, while it drops to zero at the Poplavok and at the Nizhny ponds. In summer, the distribution of amateur fishermen varies greatly: at the Forelevy and the Nizhny ponds, the values reach 2.5 persons/km, and at the Poplavok —1 person/km.

(3) Roach (53.4% of the total number in catches), crucian carp (23.2%) and perch (11.6%) predominate in the catches of amateur fishermen in the studied reservoirs as a whole. In the Forelevy pond, the main catch was roach (65.7%), perch (13.8%) and silver bream (13.0%). In the Verkhny pond, in addition to roach (73.1%), perch was often caught (25.8%). In the Nizhny pond, crucian carp (69.0%) and roach (30.1%) are more common in the catches, while in the Poplavok pond, crucian carp is usually present in the catches (87.6%).

(4) All the water bodies under consideration are eutrophic (with the exception of the mesotrophic Poplavok pond), the species abundance of the ichthyofauna of urban water bodies is small. The ponds are dominated by fish species that are undemanding to the conditions of the aquatic environment, and have a wide range of tolerance to fluctuations in environmental factors. In addition to the invasive species (sterlet, trout) grown on the farm, the Forelevy pond is also home to roach, perch, silver bream, bream, pike and eel. The representative fish species of the Verkhny Pond are perch and roach. In the Nizhny pond, the species structure is different from other water bodies — it includes not only perch and roach, but also tench and crucian carp (in the largest quantities). The Poplavok pond has a sparse ichthyofauna — it is characterized by small specimens of roach and crucian carp, which are caught by fishermen in the largest amounts. The predominance of these species is typical for mesosaprobic water bodies, which is confirmed by the data of hydrochemical monitoring.

(5) Information on the ichthyofauna in the catches of amateur fishermen can be used for a preliminary assessment of the ecological state of water bodies, giving a fairly reliable result, since the species composition and number of fish depend on water quality. Amateur fishermen are among the first to notice changes in environmental conditions in the reservoir, fish kill. Therefore, the decline in the popularity of the reservoir among amateur fishermen can be a kind of "public" indicator of water quality, and amateur fishing itself contributes to raising public awareness of the ecological state of water bodies.

References

FAO (1997) Inland fisheries. FAO Technical Guidelines for Responsible Fisheries. Food and Agriculture Organization, Rome

FAO (2018) The state of world fisheries and aquaculture 2018. Food and Agriculture Organization, Rome

Federal Law (2018) On amateur fishing and amendment of certain legal acts of the Russian federation. (December 25, 2018 No. 475-FZ). Moscow, Russia (in Russian)

Gubin AB (2006) Toponyms of Kaliningrad. Rivers and reservoirs. In V. N. Maslov (Ed.), Kaliningrad archives (pp. 64–94). KSU: Kaliningrad (in Russian)

McInnes K, Taylor SM, Webley J (2013) Social, attitudinal and motivational recreational fishing survey. Social, attitudinal and motivational recreational fishing survey: part of the 2010 statewide recreational fishing survey. Project Report. State of Queensland, Brisbane

Pravdin IF (1966) Guide to the study of fish (mainly freshwater). Food Industry, Moscow (in Russian)

Ryabtseva EA, Tsupikova NA (2019) Spatial distribution of hydrochemical parameters in the Verkhny pond (Kaliningrad). KSTU NEWS 54:61–74 (in Russian)

Shakhnazarova AB (2005) Morphophysiological and biochemical parameters of aquatic organisms under conditions of nitrite intoxication (Dissertation of Candidate of Biological Sciences). Dagestan State University, Makhachkala, Russia (in Russian)

Shibaev SV, Sokolov AV (2014) Method for the analysis of ichthyocenoses of small water bodies in the Kaliningrad region based on control catches with net fishing gear. Trudi VNIRO 151:158–164 (in Russian)

Nature Protection. Hydrosphere. Indices of state and regulations for valuation survey of fishery waters (1977) GOST 17.1.2.04–77 dated July 01, 1978. Moscow, Russia: Publishing house of standards. pp. 51–62 (in Russian)

Tsupikova NA, Sevostianova EA (2021) Some hydrological and hydrochemical peculiarities and problems of small urban ponds on the example of the Poplavok pond. KSTU NEWS 62:50–64 (in Russian)

Vasil'ev YUS, Kukushkin VA (1988) Use of reservoirs and rivers for recreation. Gidrometeoizdat, Leningrad (in Russian)

20.5 Conclusion

Thus, surveys with gill nets with different mesh sizes allow to assess the fish abundance and the allowable catch. This approach is quite objective since it is based on actual catch data and reflects the current state of fish stocks.

The method can be applied to any small water body inappropriate for fishing by gears such as a trawl or seine to calculate the fish absolute abundance.

References

Kitaev SP (2007) Fundamentals of limnology for hydrobiologists and ichthyologists. Karelian Scientific Center of the Russian Academy of Sciences, Petrozavodsk, 395 pp

Shibaev SV (2004) System approach in fisheries research. Kaliningrad, Kaliningrad State Technical University, 312 pp. https://search.rsl.ru/ru/record/01002716614. ISBN 5-94826-044-5 (in Russian)

Shibaev SV (2014) Fishery ichthyology. Mishutkina I.V., Kaliningrad, PE, 535 pp (in Russian)

Shibaev SV, Aldushin AV (2011) Spatial-temporal dynamics of ichthyocenosis of the pelagic zone of the Vistytis Lake in the Kaliningrad Region. Current status of inland waters biological resources. In: Proceedings of the First All-Russian conference with foreign partners. September 12–16, 2011, AQUAROS, Borok, Russia, 901 pp, pp 852–859 (vol 2, 468 pp). ISBN 978-5-901652-14-5 (in Russian)

Shibaev SV, Sokolov AV (2014) Method of analysis of ichthyocenosis in small water bodies of the Kaliningrad oblast by means of catches of gill nets. TRUDY VNIRO 151:158–164. http://dspace.vniro.ru/handle/123456789/7301 (in Russian)

Shibaev SV et al (2008) Current status of research on the morphological features of the Vistytis Lake. In: Shibaev SV, Bernikova TA, Aldushin AV, Lyach DM (eds) Innovation in the science and education, 2008: Proceedings of VI international scientific conference (21–23 October 2008) Part 1. FGBOU VPO "KSTU", Kaliningrad, pp 156–158 (in Russian)

Shibaev SV et al (2017) Current status of the Lake Vistytis in the Kaliningrad Region. In: Gritsenko VA, Sivkov VV, Yurov AV, Kostianoy AG (eds) Terrestrial and inland water environment of the Kaliningrad Region. The handbook of environmental chemistry, vol 65. Springer, Cham. https://doi.org/10.1007/698_2017_112

Vistytis Lake (2008) In: Tylik KV, Shibaev SV (eds). Mishutkina I.V., Kaliningrad, PE, 144 pp. ISBN 978-5-98787-047-1 (in Russian)

Part III
Fisheries Management

Chapter 9
Prospects of Development of Small-Scale Fisheries of Vendace in Lake Vistytis, Kaliningrad Oblast, Russia

A. V. Aldushin and S. V. Shibaev

Abstract The purpose of this work was to assess the stock of one of the most abundant target of fisheries in Lake Vistytis, the European vendace, as well as the prospects for its fisheries. The pelagic ichthyocenosis of the lake and the role of vendace in its formation were studied based on three main sources of information, the data of net catches, experimental trawling and hydroacoustics. The species structure was determined based on the data of control catches using set gillnets with a mesh size of 6.5–40 mm and a pelagic trawl with a 5 mm codend mesh size. Hydroacoustic surveys were aimed at studying the spatial distribution of fish in the water column, estimating their absolute abundance, as well as restoring the size structure based on the dependence of the size of the fish on the target strength or acoustic size. The results of the control catches showed that vendace is the main component of the pelagic ichthyocenosis. Experimental surveys with a pelagic trawl made it possible not only to study the possibilities of using this fishing gear in the water body, but also to catch vendace fingerlings for the first time and to take into account their biological parameters when assessing the prospects for the rational use of this species. The results of hydroacoustic studies made it possible to clearly demonstrate the dependence of the spatial distribution of vendace on the temperature, oxygen and morphological conditions in the lake, as well as to determine the optimal fishing areas. The assessment of the abundance and biomass showed that the stocks of this species are at a good level, and the potential yield is comparable to that in Soviet times during the period of intensive fishing. The almost complete absence of other fish species in the catches in the pelagic area of the lake allowed us to conclude that the effect of the specialized fishing of vendace on other representatives of the ichthyofauna is insignificant, as well as to assess the prospects for its fishing with both passive and active fishing gear.

Keywords Lake vistytis · Vendace · Pelagic Ichthyocenosis · Hydroacoustic estimate · Abundance · Biomass · Potential yield first section

A. V. Aldushin (✉) · S. V. Shibaev
Kaliningrad State Technical University, Sovetsky avenue, 1, 236022 Kaliningrad, Russia
e-mail: aldushin@klgtu.ru

A. G. Arkhipov (ed.), *Sustainable Fisheries and Aquaculture: Challenges and Prospects for the Blue Bioeconomy*, Environmental Science and Engineering, https://doi.org/10.1007/978-3-031-08284-9_9

9.1 Introduction

The organization of sustainable small-scale fisheries, which currently receives a lot of attention (Voluntary guidelines 2015) should be based on solving three issues: (1) determining promising stocks for fisheries, (2) assessing biologically safe management guidelines, including the concept of an ecosystem approach, and (3) finding optimal parameters of fisheries from a technical and economic point of view. At the same time, solving these issues is significantly difficult in small water bodies, because due to the low total fish productivity, it is not appropriate to conduct their large-scale scientific studies. This paper presents approaches to address these tasks with regard to the organization of small-scale vendace fisheries in Lake Vistytis.

Lake Vistytis is located in the southeast of Kaliningrad oblast at an altitude of 172 m above sea level, has a glacial origin with a complex bottom relief. According to hydrochemical parameters it is an oligotrophic water body with some features of mesotrophy in the coastal areas. The maximum depth is 54 m; the surface area is 1.8 ha. The depth of almost 2/3 of the total area of the lake is up to 20 m, and only about 13% of the area are deeper than 30 m (Shibaev et al. 2017).

Before 1990, vendace (*Coregonus albula* (L, 1758)) was the main target of local fisheries in Lake Vistytis, and its maximum production reached 29 tons, averaging about 16 tons per year (Tylik et al. 2008). Fishing was carried out using gillnets with a 18-mm mesh size. Due to decreasing commercial fisheries in the 1990s, vendace fishing practically stopped, and attempts to restore it in the last decade were ineffective. The studies conducted in 2007–2009 showed that the size structure of the vendace population changed towards a decreasing length (Shibaev et al. 2009). The reason for a decreasing length of this species could be either overpopulation due to the absence of fishing activity for a 20-year period, or an increase in natural mortality of fish due to the high prevalence of infection with the parasites *Proteocephalus exiguus* and *Ergasilus briani* (Evdokimova et al. 2019). The latter may also be due overpopulation, which increases the spread rates of parasites. The change in the size and age structure of the vendace population resulted in a reduction of the efficiency of fishing with traditional fishing gear, set gillnets with a 18-mm mesh size: catches per unit effort turned out to be so low that, compared to the 1970–1980s, they made fishing unprofitable. At the same time, catches with nets with a smaller mesh size, 12–16 mm, were by an order of magnitude higher, which made it possible to use them in fisheries (Shibaev et al. 2009) (see Fig. 9.1).

This conclusion is confirmed by the results of studies aimed at estimating the absolute number of vendace by the method of hydroacoustics (Aldushin and Shibaev 2016): the average annual abundance of the species is, in general, at a relatively high level, 12.1–15.8 million ind. with the biomass (estimated by weighted average length and weight) of 84–112 tons (Aldushin et al. 2016), which can ensure a catch commensurate with the values achieved during a period of good fisheries management. The purpose of this study is to assess the optimal regime of vendace fisheries in Lake Vistytis in modern conditions.

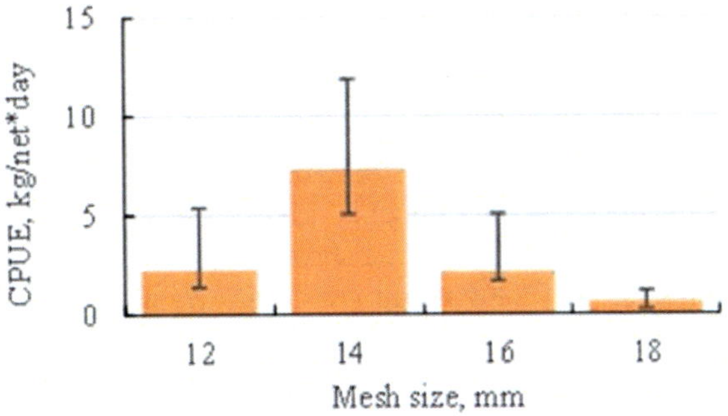

Fig. 9.1 Catches per unit effort of vendace in lake vistytis with set gillnets with a mesh size of 12, 14, 16 and 18 mm (our data)

9.2 Materials and Methods

This work is based on comprehensive fisheries studies of the ichthyofauna in Lake Vistytis for the period 2007–2021. In the study of pelagic ichthyocenosis in general and vendace in particular, three main sources of information were used: (1) data of gillnet catches, (2) experimental trawling and (3) hydroacoustics.

The species structure of the pelagic ichthyocenosis in Lake Vistytis was determined on the basis of the data of the control catches using set gillnets with a mesh size of 6.5, 9, 10, 12, 14, 16, 18, 20, 22, 24, 35 and 40 mm. Gillnets were set in the water column every 5 m, starting from the surface, in all horizons at a distance of at least three meters from the bottom in order to avoid fishing in the bottom horizon. The fishing lasted 0.5–1 day.

Trawling surveys were carried out with a pelagic trawl in the daytime and at night in the direction from the central deep-water depression of the lake to the shore of its northern part. The fishing horizon was set by selecting the required length of warps and was determined by the AsCor hydroacoustic complex (Degtev and Ivanter 2002). The design of the trawl was characterized by the following parameters: pelagic (allopelagic); four-layer; with mesh sizes of 80, 60, 30, 18, 12, 10, 5 mm; its total length was 40 m. A trawl was towed by two Oka-4 boats (weight of 190 kg) (pair trawling) with outboard motors of 15 hp. The distance between the boats ranged from 25 to 60 m, depending on the warp length; vertical opening was 5 m; horizontal opening was 7 m; the trawling speed was 3–5 km/h (Shibaev et al. 2012).

Hydroacoustic surveys were carried out in the spring–summer period during the daytime and at night using the AsCor hydroacoustic complex based on the Furuno LS-4100 echo sounder according to the meander type tack configuration (Aldushin et al. 2016). The restoration of the size composition of the detected hydrobionts was carried out based of the regression equation given in the literature (for a frequency of 120 kHz) for vendace (see formula (9.1)) (Mehner 2006), taking into account the correction for frequency (see formula (9.2)):

$$TS = 25.5LgL - 70.9 \tag{9.1}$$

$$d = 0.9Lg\frac{F}{f} \quad (9.2)$$

where TS is the target strength, dB; L is the length, cm; F is the frequency used, kHz; f is the reference frequency, kHz.

The calculation of the absolute number of fish in Lake Vistytis was made out on the basis of the Kriging geostatistical interpolation method (Rivorad 2000) with a linear variogram model and 100 m grid nodes. The number of fish was calculated as the volume of a three-dimensional figure (Borisenko 2013). The reflected signals from fish with a target strength value (TS) of -50 dB < TS < -25 dB were taken into account when estimating the number of fish by the hydroacoustic method, and the target strength values of -60 dB < TS < -25 dB were taken into account when studying the dimensional structure.

All materials on the control catches are standardized and reduced to general parameters: the frequency of occurrence of fish species in fishing gear, the catch per unit effort CPUE (catch per day of fishing with a net of 25-m long for gillnets, catch per hour of trawling for trawl), expressed numerically (Yn/f) and in weight (Yw/f). Considering that, all other things being equal, the catch per effort is determined by the density of the fish community, these parameters can be considered as indices of the abundance and biomass of a particular species, taking into account the selectivity of the net with a certain mesh size.

9.3 Results

Surveys of pelagic ichthyocenosis in Lake Vistytis using gillnets and a pelagic trawl have showed that vendace is its dominant component; it was recorded in more than 70% of catches, and accounted for more than 97% in terms of abundance and more than 95% in terms of biomass in catches. The maximum abundance index was recorded in set gillnets with a mesh size of 9–14 mm, and the maximum abundance and biomass were obtained with nets of 9–14 mesh size for vendace. The biomass of other species increased in fishing nets with a larger mesh size (35, 40 mm); however, the catches of such species (mainly perch, roach, and pike) in terms of abundance were rare. A similar pattern was observed in trawl catches, in which the size groups from 3 to 14 cm were mainly represented by vendace, and the size group of 11–13 cm had the largest biomass values (see Fig. 9.2). In addition, trawl surveys made it possible for the first time to obtain catches of vendace fingerlings and take into account their biological parameters when assessing the prospects for the rational use of this species.

As for the vertical distribution, vendace was represented in almost all fishing horizons and, regardless of the season, it accounted for the largest catch values, with the exception of the epilimnion (0–10 m), where other fish species (bleak, roach,

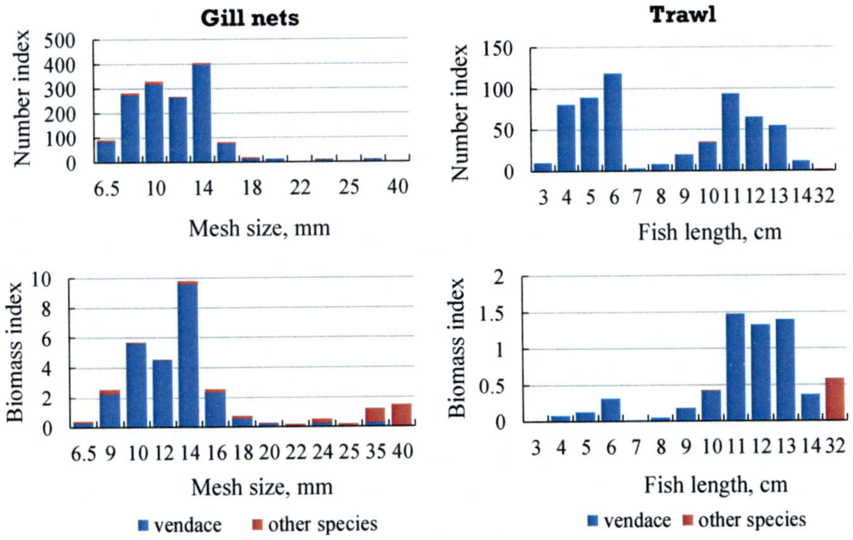

Fig. 9.2 Size and species composition of the pelagic ichthyocenosis in lake vistytis based on the data of control catches (our data)

perch, pike) prevailed in catches. This fact made it possible to apply the hydroacoustic technique to study the spatial distribution of the species and to determine its abundance and biomass. The results of the surveys allowed us to clearly demonstrate the dependence of seasonal vertical migrations of vendace on the thermal conditions in the lake. Thus, during the spring and autumn homothermy, at night, vendace is distributed throughout the water column in a layer of 10–35 m deep, the height of the layer of aggregations can reach 25 m, and the maximum aggregations are formed in a layer of 15–25 m deep. In summer, during the formation of the temperature jump layer, vendace concentrates under the thermocline in the deep-water zone at night and it does not ascend above 12–15 m. Two main layers of aggregations are distinguished, 15–20 and 25–35 m. The main aggregations are formed in a layer of 25–35 m. A similar pattern of the vertical distribution of vendace is found in net catches in these periods (see Fig. 9.3).

The distribution of vendace in the water area of the lake is determined by its temperature and morphological conditions. During the spring homothermy, the density map of its aggregations has a dispersed pattern of distribution over the water area, starting from depths of 12–15 m. As the temperature jump layer is formed (summer period), the water area acceptable for its habitation is reduced by more than 1.5 times, vendace migrates towards larger depths (15 m and below), and its aggregations are confined to deep-sea depressions (see Fig. 9.4).

The results of the study of the spatial and temporal dynamics of vendace in Lake Vistytis are important both for estimating the abundance of this species and assessing the optimal areas and periods of its fisheries. It is also important that pelagic ichthyocenosis (in the depth range from 10 m to the bottom layer) is represented

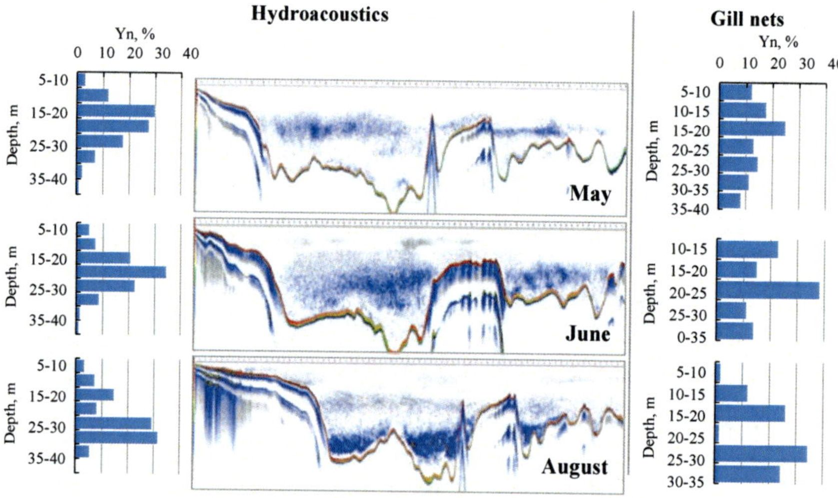

Fig. 9.3 Seasonal dynamics of vertical migrations of vendace (our data)

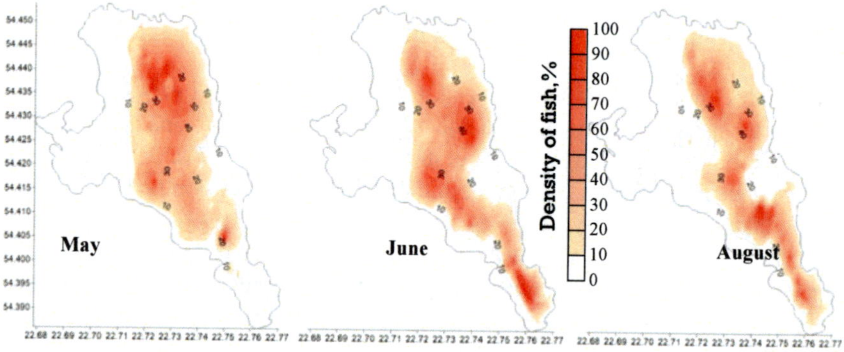

Fig. 9.4 Seasonal dynamics of distribution of vendace in the lake area based on the data of hydroacoustics (our data)

exclusively by vendace, which suggests that its specialized fisheries in the water column with both passive and active fishing gear will not have a significant impact on other fish species. This fact makes it possible to assess the prospects for vendace fisheries both with traditional set gillnets and with a pelagic trawl.

The abundance of vendace was estimated using the Kriging geostatistical modeling. The results of a comparative analysis of the dynamics of changes in the abundance according to the data of hydroacoustic surveys and catches per effort with set gillnets make it possible to conclude that there is a direct relationship between these parameters, which indicates the reliability of the data resulted from the hydroacoustic surveys and makes it possible to use them for the estimation of the species abundance (see Fig. 9.5)

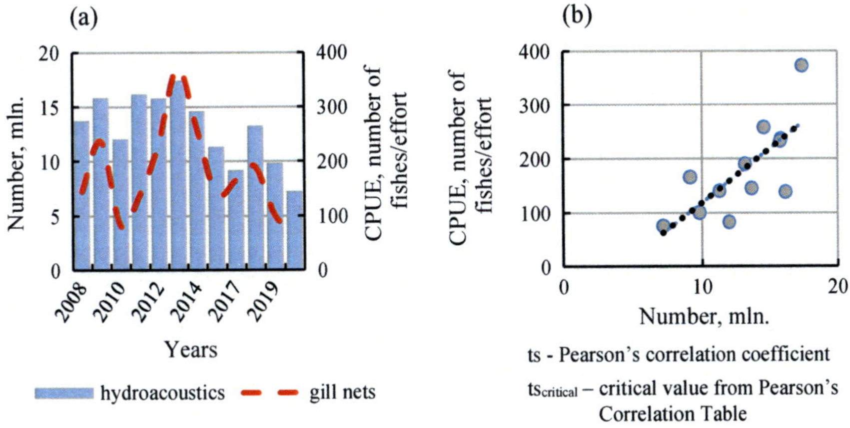

Fig. 9.5 Dynamics of changes in the abundance (hydroacoustics) and catches per effort (set gillnets) (A) and correlation between these two parameters (B) (our data)

In general, the estimation of the vendace biomass can be reduced to comparing the number of each size group and the biomass value correlated with it (in accordance with the size of the specimen). The reconstruction of the size structure of vendace was carried out based on the relationship between the fish length and the target strength according to the regression equation for vendace described in the literature (Mehner 2006). The comparison of the size structures of vendace using three main sources of information (nets, trawl, hydroacoustics) gives the following results. Due to the effect of the catchability coefficient on the size structure of vendace in gillnet and trawl catches, the variation of the abundance of vendace specimens of different sizes occurs in different ways and differs from the size structure according to hydroacoustics, however, the range of values is similar and was from 3 to 16 cm. Vendace of a length of less than 3 cm was not found both in the results of sonar surveys and in trawl catches, where the minimum recorded length was 3 cm, and specimens with a length of more than 16 cm were also practically not recorded on echogram records (the number according to sonar data was less than 1%). A similar result was also obtained in control catches with multi-mesh gillnets with a mesh size of 6.5–24 mm (see Fig. 9.6).

The dependence of the weight of vendace on its length determined according to the biological analysis (see Fig. 9.7) made it possible to use the biomass instead of the abundance of this species by comparing the size structure of vendace in Lake Vistytis, restored by the regression equation, with the number of this species, determined by the data of hydroacoustics. Thus, after 2020, when the minimum values were recorded, the abundance and biomass of vendace in 2021 increased to the level of 2019 and amounted to 9.76 million ind. and 57.8 tons (SSB = 32.4 tons) respectively (see Fig. 9.7).

Based on the concept of a precautionary approach, which determines the minimum biomass of the spawning stock equal to half of the maximum value for the virgin

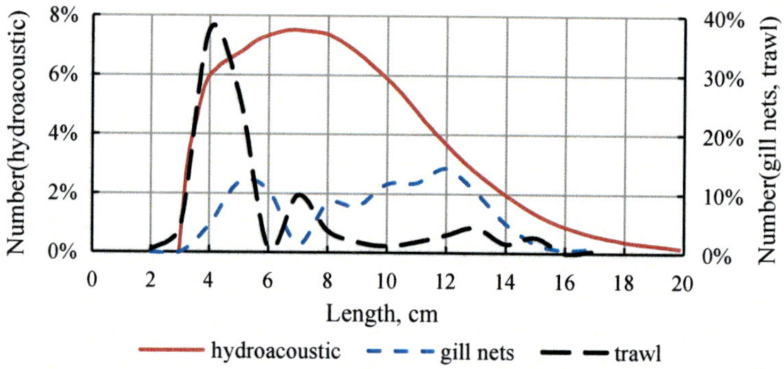

Fig. 9.6 Size structure of vendace in lake vistytis based on the data of catches and hydroacoustis (our data)

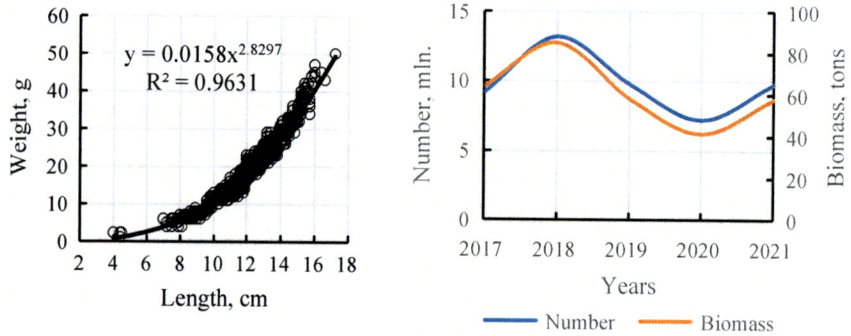

Fig. 9.7 Dependence of the weight of vendace in lake vistytis on its length (left) and dynamics of changes in the abundance and biomass of vendace based on the data of hydroacoustic surveys (right) (our data)

population (SSB50%), the yield can be about 16 tons, which is comparable to the catch volumes obtained in Soviet times (the average annual yield of vendace was about 16 tons, the maximum, 29 tons (Tylik et al. 2008). However, the catches of this species by commercial fishing enterprises are much lower. The main reason is the low fishing intensity. According to the fisheries statistics provided by fishing enterprises or fishermen engaged in fishery in Lake Vistytis to the West-Baltic Territorial Department of the Federal Agency for Fisheries, two fishing enterprises carry out commercial fishing in the lake and fishing is conducted from March to December ("Information on…", n.d.). The vendace fishery is currently conducted by one fishing enterprise, mainly from June 11 to October 31, so the period of vendace fishing is 143 days. Over the past 7 years, the catch volumes of vendace harvested by fishing enterprises do not exceed 2 tons (the quota utilization is only 10%). First of all, this is due to the fact that the vendace fishery is actually conducted by one fishing brigade, which is not able to intensify fishing. The use of gillnets with a smaller mesh size

(14 mm), showing larger catches per effort compared to traditional nets (see Fig. 9.1), in practice have shown their low efficiency in terms of the ratio between the volume of the catch and the time spent on the catch processing (fish removing and sorting) due to the increasing difficulty of untangling the fish from the net; in that regard, in fact, nets with a 16–18-mm mesh size are used in vendace fisheries.

In order to estimate the optimal parameters of the vendace fisheries, calculations were made both for passive and active fishing gear. So, in order to estimate the number of nets (the length of the gang of nets) that will be set every day to achieve a potential yield of vendace, appropriate calculations were carried out based on the average catches per effort with set gillnets with different mesh size, the fishing period in the lake (143 days), as well as the values of the potential yield (16 tons) (Table). Thus, for example, an average catch per effort with nets with a 16-mm mesh size is 2.3 kg. Accordingly, it will be necessary to set 48 gillnets per day (more than 1 km of nets) to achieve a catch of 16 tons for 143 days. As evidenced by our experience of catch unloading during ichthyological surveys in Lake Vistytis, it takes about 15–25 min to remove the fish caught from one net of a standard length (if the catch is sufficient), in the case of small-mesh (14 mm) nets, the time can increase to 25–35 min. Accordingly, it will take 12–20 h to unload the fish caught when setting 48 nets per day with a mesh size of 16 mm; when setting 15 nets per day with a mesh size of 14 mm with the maximum catch per effort, it will take 7–9 h to unload and sort the catch. At least two hours will be additionally required to set a gang of nets with a total length of about 400 m, including the travel time to the sites of the main aggregations of vendace and back, and the time spent on setting a fishing gear. Taking into account that the fishing brigade is engaged in multi-species fishing, and it also takes time to unload the catches of other nets, achieving these parameters in practice is unrealistic.

In this regard, the use of a pelagic trawl for specialized vendace fisheries in Lake Vistytis may be promising. At the same time, the absence of a specialized fleet implies the use of a small-sized trawl by boats with low-power outboard motors (15 hp) (the possibility of such vendace fisheries using pair trawling with a pelagic trawl is shown in the work (Shibaev et al. 2012)). The period (July–September) should be considered a probable time frame of trawling, when, as a rule, a layer of temperature jump has already been formed in the lake at a depth of more than 10 m, and vendace stay in the deep parts of the lake area. This period is also characterized by the presence of vendace aggregations in the pelagic zone of the lake during the daytime (mainly July–September months), when the oxygen conditions in the lake (low content of dissolved oxygen) affect the lower boundary of its vertical distribution. The most optimal course of trawling with sufficient depths and the largest length in a straight-line direction may be the course from the central deep-water depression to the northern part of the lake coastline (see Fig. 9.8), where the largest aggregations of vendace are concentrated.

Taking into account the results of experimental trawling (Shibaev et al. 2012), when catches per effort (kg/h of trawling) in the daytime were about 60 kg/h, achieving a catch value of 16 tons will require 3 h of trawling per day for 90 days (Table). At a vessel speed of 3 km/h (which was reached during experimental trawling), one trawling operation will take almost an hour, plus the time spent on

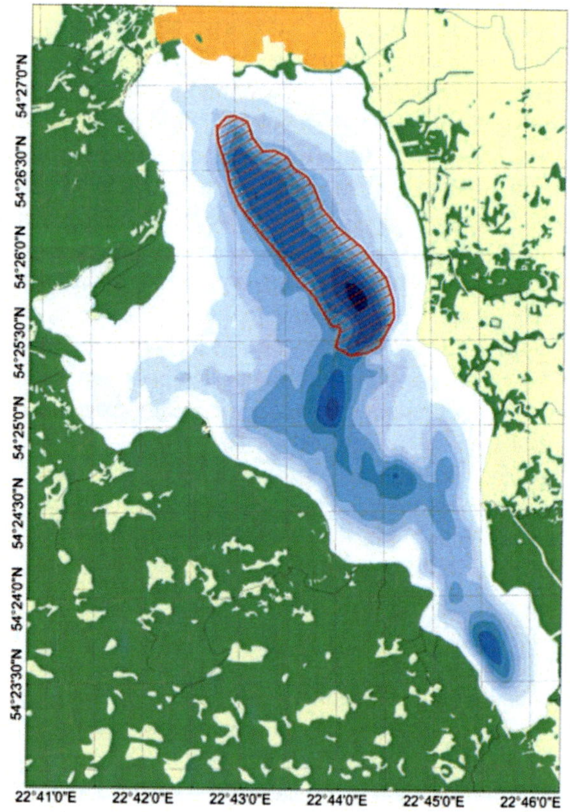

Fig. 9.8 Optimal region of trawling (our data)

retrieving and unloading the trawl, transporting to the shore, sorting the catch, and preparing the trawl for work, which will be at least an hour and a half and in general, it takes of about 9 h of the working time. A decrease in the period of trawl fisheries to 90 days will make it possible to maintain the multi-species fisheries of a fishing brigade, redistributing it over time: in the summer months, during the period of the highest tourist activity in the lake, this will allow focusing on the target fishing of vendace, a valuable species in terms of taste in Lake Vistytis, the maximum aggregations of which are formed in the water column and, and the rest of the time pursue fishing of other fish species.

9.4 Conclusion

The studies have shown a good current state of the vendace stock in Lake Vistytis, and allow us to conclude about the possibility of organizing sustainable small-scale fisheries of the species in the lake. The vendace fisheries using set gillnets with a 16–18-mm mesh size, currently used for its fishing, shows their low efficiency due

Table 9.1 Calculation of the required number of nets set per day/trawling per hour to achieve a potential yield

Fishing parameters	Set gillnet				Pelagic trawl
	Mesh size, mm				Codend mesh size, mm
	12	14	16	18	5
Catch per effort (kg/25-m net per day)	2.5	7.3	2.3	0.6	60
Fishing period, days	143				90
Possible yield, t ($SSB_{50\%}$)	16				
Number of nets per day / trawling per hour	45	15	48	187	3
Length of a gang of nets, m	1125	375	1200	4675	-

to smaller sizes of specimens (probably due to overpopulation), and nets with a 14-mm mesh size significantly reduce the commercial quality of fish and increase the time of the catch unloading. One of the solutions to this problem and increase in the catches of vendace within the framework of the development of fisheries in the lake could be the introduction of trawl fishing of the species. This will improve its marketable quality, preserve the multi-species fisheries in the lake redistributing it over time, and the increase in the production of vendace will increase its growth rate due to higher food availability. The use of a pelagic trawl for vendace fisheries will not have a significant impact on other representatives of the ichthyofauna in the lake. All these factors make it possible to conclude about compliance with the principles of the ecosystem approach and ensuring sustainable fishing in Lake Vistytis (Table 9.1).

References

Aldushin AV, Shibaev SV (2016) Hydroacoustic studies of pelagic ichthyocenosis in Lake Vistytis in Kaliningrad Oblast. Inland Water Biology 4:54–63 (in Russian)

Borisenko ES (2013) Hydroacoustic studies of the distribution of fish in the floodplain-channel system of the Lower Irtysh. Dissertation of Candidate of Biological Sciences. IPEE RAN, Moscow, Russia (in Russian)

Degtev AI, Ivanter DE (2002) Automated system for the quantitative assessment of fish stocks by the hydroacoustic method ASKOR-2. Fisheries 4:32–40 (in Russian)

Evdokimova EB, Arkhipovskiy SN, Zaostrovtseva SK, Avdeeva EV (2019) Parasitofauna of the vendace (Coregonus Albula Linnaeus, 1759) of Lake Vistytis in 2016 and 2018. In: Kostrikova NA (eds.) VII Baltic Marine Forum. Kaliningrad: KSTU (in Russian)

FAO (2015) Voluntary guidelines for securing sustainable small-scale fisheries in the context of food security and poverty eradication, Rome

Information on the development of allocated quotas (n.d.). Retrieved from http://zbtu39.ru/informacziya-ob-osvoenii-vydelennyh-kvot/(inRussian)

Mehner T (2006) Prediction of hydroacoustic target strength of vendace (Coregonus albula) from concurrent trawl catches. Fish Res 79:162–169 (in Russian)

Rivoirard J, Simmonds J, Foote KG, Fernandes P, Bez N (2000) Geostatistics for estimating fish abundance. Blackwell Scientific Publications, Great Britain

Shibaev SV, Sokolov AV, Aldushin AV (2009) To optimization of vendace fishery (Coregonus albula L.) in the lake vistytis of the kaliningrad oblast in modern conditions. In: Volkogon VA and Ivanov AV (Eds.). Innovations in science and education—2009: VII International science conference. Kaliningrad: KSTU (in Russian)

Shibaev SV, Levchenko SV, Aldushin AV, Novozhilov OA, Belykh AV (2012) The first results of experimental trawl fishing in the lake Vistytis in Kaliningrad Oblast. In V.A. Volkogon and A.V. Ivanov (Eds.). Innovations in science, education and business: IX International scientific conference. Kaliningrad: KSTU (in Russian)

Shibaev SV, Sokolov AV, Tylik KV, Bernikova TA, Shibaeva MN, Masyutkina EA, Zaostrovtseva SK (2017) Current status of the lake Vistytis in Kaliningrad region. Terrestrial and Inland Water Environment of the Kaliningrad Region 4:441–480 (in Russian)

Tylik KV, Sokolov AV (2008) Fishes of lake vistytis. Kaliningrad, Russia: IP Mishutkina (in Russian)

Chapter 10
Rational Use of Aquatic Biological Resources of the Atlantic Ocean for the Sustainable Economic Development of Russia

A. G. Arkhipov, V. I. Sauskan, and V. M. Osadchiy

Abstract The concept of the "blue economy" reflects the recognition of the exceptional role of the oceans and seas for the world community. The rapid degradation of aquatic ecosystems and the depletion of their resource base due to an increase in environmental and economic burden inevitably have an impact on the well-being and health of people living in coastal countries. With the increase in the scale of oceanic fishing, there was a growing understanding of the need to study and preserve the living resources of the oceans. This was facilitated by a number of objective factors and, above all, the desire of all States involved in the exploitation of aquatic biological resources to create conditions for the sustainable, maximum possible, rational use of these resources. In order to preserve living marine resources for most areas of the world Ocean, including in the Atlantic, international legal acts have been prepared and signed. As a result, international fishing organizations were created for almost every district: fishing commissions, scientific councils, scientific committees, etc. The main task for our country in the research and extraction of aquatic biological resources of the Atlantic Ocean in modern conditions is to protect the interests of optimal Russian fishing at the international level. This is possible by evaluating aquatic biological resources on a strict scientific basis and developing recommendations for their rational use, taking into account the Concept of the "blue economy", the requirements of international conventions and bilateral treaties. In total, in the oceanic regions of the Atlantic, the domestic fleet can annually extract about 1.0 million tons of aquatic biological resources without damage to the oceanic ecosystem. In addition, it should be borne in mind that there is a promising reserve—possible fishing of Antarctic krill up to 5.5 million tons annually.

Keywords The concept of the "blue economy" · The Atlantic ocean · Oceanic fishing · Aquatic bioresources

A. G. Arkhipov · V. I. Sauskan (✉) · V. M. Osadchiy
Kaliningrad State Technical University, Kaliningrad, Russia
e-mail: sauskan@klgtu.ru

A. G. Arkhipov
e-mail: ag_arkhipov@mail.ru

A. G. Arkhipov (ed.), *Sustainable Fisheries and Aquaculture: Challenges and Prospects for the Blue Bioeconomy*, Environmental Science and Engineering, https://doi.org/10.1007/978-3-031-08284-9_10

10.1 Introduction

The problems of rational use of aquatic biological resources (ABR) within the framework of the "blue economy" Concept are given great attention in international fisheries management organizations, as well as in the formation of national development strategies of coastal countries. The peculiarity of this Concept is the emphasis on the systematic and sustainable development of economic sectors related to the use of the resources of the World Ocean. The development of the Concept reflects the recognition of the exceptional role of oceans and seas for the world community. The rapid degradation of aquatic ecosystems and the depletion of their resource base due to an increase in environmental and economic burden inevitably have an impact on the well-being and health of people living in coastal countries.

Currently, the term "blue economy" is widely used around the world and has three interrelated, but somewhat different meanings: the overall contribution of the oceans to the economy; the need to address issues of environmental sustainability of the oceans; the development of the ocean economy as a growth opportunity for both developed and developing countries. Despite the growing challenges, the World Ocean still provides humanity with huge opportunities for the development of various resources, including the ABR. A general consensus on understanding the concept of the "blue economy" and using the potential of the oceans for growth was first reached by the countries at the UN Conference on Sustainable Development "Rio + 20" in 2012. In 2015, the idea of the concept formed the basis for the definition of the UN Sustainable Development Goals, in particular, the 14th goal sounds like "conservation and rational use of oceans, seas and marine resources for sustainable development". The concept of the "blue economy" is aimed at the systematic and sustainable development of economic sectors related to the use of the resources of the World Ocean, and reflects the desire of the world community to find a balance between further economic growth and development, preservation of environmental safety and ensuring the social well-being of the world's population (UN Resolution 2015; https://www.kp.ru /; https://ec.europa.eu).

The purpose of this work is to carry out a brief analysis of the current state of aquatic biological resources in the fishing areas of the Atlantic, and to assess the scientifically sound possibilities of the extraction of ABR by the domestic fishing fleet, which do not contradict the Concept of the "blue economy" in these areas.

10.2 Results and Discussion

With the increase in the scale of oceanic fishing, there was a growing understanding of the need to study and preserve the living resources of the oceans. This was facilitated by a number of objective factors and, above all, the desire of all States involved in the exploitation of living biological resources to create conditions for the sustainable, maximum possible, rational use of these resources. Therefore, in 1958, at the first UN

Conference on the Law of the Sea, the Convention "On Fisheries and the Protection of Living Resources of the High Seas" was adopted. The signatory States have committed themselves to jointly develop measures for the conservation and rational use of the living resources of the oceans.

To manage the stocks of major commercial fish, the Food and Agriculture Organization of the United Nations (FAO) has identified fishing areas for all oceans, including seven areas for the Atlantic Ocean (see Fig. 10.1).

In order to preserve living marine resources, international legal acts have been prepared and signed for most areas of the Atlantic. As a result, international fishing organizations were created for each area: fishing commissions, scientific councils, scientific committees, etc. In all the adopted international legal acts in the field of fisheries, much attention was paid to marine scientific research. At the III UN

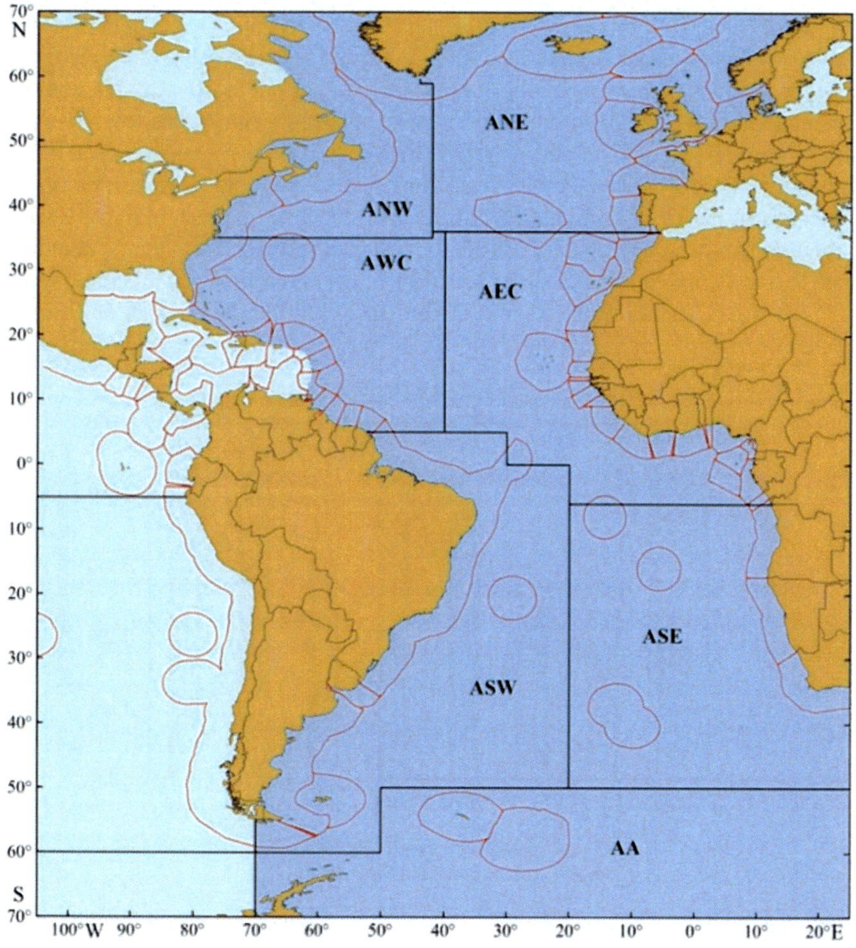

Fig. 10.1 Fishing areas of the Atlantic ocean (zoning according to FAO)

Conference on the Law of the Sea in 1982, the "UN Convention on the Law of the Sea" was adopted. Considering that the habitats of many commercial species in the World Ocean are located both in exclusive economic zones and in the open sea, the problem of conservation of straddling and highly migratory fish was considered at special UN conferences in 1993–1995. At the final conference in 1995, the Agreement "On the Implementation of the Provisions of the 1982 UN Convention" was adopted, which concerns the conservation and management of Straddling Fish Stocks, Highly Migratory Fish Stocks. Coastal States and States fishing on the high seas have become obliged to take measures aimed at the sustainable existence of straddling fish stocks and highly migratory fish stocks, and to promote the goal of their optimal use. At the same time, reliable scientific data should be used (Sauskan et al. 2020a, b).

Soviet/Russian scientists and specialists have been actively involved in all stages of creating an international legal framework for the conservation of living resources of the World ocean and studying their condition. Thanks to the large-scale scientific research carried out by our country, primarily in the Atlantic Ocean, measures have been developed and taken aimed at the long-term and sustainable use of the reserves of the most valuable types of aquatic biological resources of the Atlantic. The activities of international fisheries commissions in most of the convention areas of the Atlantic were based almost until the end of the twentieth century on the results of expeditionary research of the Soviet scientific fleet. This made it possible to ensure the fishing interests of our country in this part of the World Ocean, to create reliable working conditions for the fishing fleet. Currently, fisheries research the Atlantic continues, but to a lesser extent. The catch of aquatic biological resources by all countries and the catch of Russia in the Atlantic Ocean (including the Norwegian and Barents Seas) in recent years is shown in Fig. 10.2. The world catch was fairly stable and

Fig. 10.2 World catch of aquatic biological resources and catch of Russia in the Atlantic ocean

ranged from 19–22 million tons, the catch of Russia varied from 1.1 to 1.5 million tons (FAO Yearbook 2016, 2020; http://fish.gov.ru).

Next, we will consider the directions of research and the volume of catch of ABR in recent years by all countries and Russia in seven regions of the Atlantic (Bandurin and Arkhipov 2017; Sauskan et al. 2018).

Northwest Atlantic (ANW). Bioresource fishing is regulated by the Northwest Atlantic Ocean Fisheries Organization (NAFO). Russian specialists monitor the domestic fishery and periodically participate in surveys to assess the stocks of sea bass. The total catch of ABR by all countries fishing in this area ranges from 1.8–2.1 million tons. The recommended catch of Russia is approximately 5.0 thousand tons. For the Russian fishery in the ANW, sea bass, Greenland halibut and beryx are of the greatest importance. Currently, the possibilities of Russian fishing in this area are relatively small. The total volume of quotas allocated to Russia in recent years was up to 20 thousand tons (without a quota for catching pelagic perch, this resource was distributed jointly for the ANW and ANE).

Northeast Atlantic (ANE). Very effective monitoring of fishing is carried out in the region within the framework of intergovernmental agreements and international regional organizations. As in the ANE, domestic specialists monitor the Russian fishery and participate in international surveys to assess the stocks of sea bass. The possible catch of ABR in open waters is distributed among the member countries of the Northeast Atlantic Fisheries Commission (NEAFC). The annual total global catches of aquatic biological resources in recent years have been at the level of 8.0–9.3 million tons. The raw material base of Russian fishing in the ANE is represented by pelagic species—whiting, haddock and sea bass. Total in ANE (without catches in the Barents and Norwegian Seas) Russia can produce up to 80–90 thousand tons.

Central-Eastern Atlantic (AEC). The basis of the raw material base of the fishery in AEC is the stocks of mass pelagic fish—European sardine, European and West African horse mackerel, caranx, round and flat sardinella, Eastern mackerel and some other species. The increase in catch is possible mainly due to sardines, the number of which remains at a stable level, and to some extent due to mackerel. The total catch of ABR by all the countries fishing in this area has been quite stable in recent years and ranged from 4.2–5.1 million tons. Monitoring of fishing in AEC is carried out by the Food and Agriculture Organization of the United Nations (FAO). Russian specialists monitor the domestic fishery and conduct annual surveys to assess the replenishment and stocks of mass pelagic fish. AEC is a traditional area of domestic fishing. Currently, Russia's catch in AEC (in the Exclusive economic zones of foreign countries) is at the level of 140–240 thousand tons/year. The annual catch of Russia in this area (according to intergovernmental agreements in the Exclusive economic zones of foreign states) is possible in the amount of 250–260 thousand tons.

Central-Western Atlantic (AWC). The biological resources of the AWC are represented by pelagic (horse mackerel, herring, sardinella, tuna, marlin, swordfish, mackerel, mackerel, sharks) and demersal (merow, lutyans, asparagus, pristipomovye, mullets, sea burbot, sharks, rays, etc.) fish, as well as invertebrates (shrimp, crabs, lobsters, squid, bivalves and gastropods). The world catch in this area ranges from 1.2–1.5 million tons. The Russian fishing fleet has not worked in the AWC in recent decades. There have also been no studies of ABR reserves by domestic specialists. Fishing statistics from available international sources are analyzed.

Southeast Atlantic (ASE). The stocks of the main commercial objects of this area—horse mackerel, sardinella and hake, are in satisfactory condition. The total catches of aquatic biological resources in ASE in recent years have varied in the range of 1.3–1.7 million tons. Currently, the countries of the region independently regulate the exploitation of reserves, coordinating their efforts within the framework of the Bengal Current Commission established by them. Russian specialists have not conducted any studies of the ABR reserves in this area in recent decades. Fishing statistics from available international sources are analyzed. In recent years, Russia has extracted up to 15.0 thousand tons of fish in ASE by agreement with enterprises of Angola and Namibia. Annually in ASE, Russia, within the framework of joint ventures with coastal countries, can produce 250–260 thousand tons of fish.

The Southwest Atlantic (ASW) is one of the most important areas of the world's fisheries. The annual total catches of seafood in the last decade were at the level of 1.6–2.4 million tons. The fishing of bioresources in the ASW has been stable these years with a catch close to 2 million tons. Catches above this value, as a rule, are obtained only in years of very high catches of Argentine squid. In many ways, it is the fluctuations in squid catches that determine the variability of the total catches of aquatic biological resources in this area. There have been no studies of ABR reserves by domestic specialists and Russian fishing in recent years in the ASW. Fishing statistics from available international sources are analyzed. The total catch of Russia in this area, mainly due to the extraction of squid, may amount to about 40 thousand tons.

Antarctic part of the Atlantic (AA). Antarctic krill is the main commercial resource in the waters of the Southern Ocean. According to the combination of catch potential and consumer properties, Antarctic krill is currently the largest and most promising resource of the World Ocean. In addition, fangfish and ice fish are fished in AA. The dynamics of the total catch of commercial AA hydrobionts (primarily krill) varied within 0.16–0.30 million tons. All modern krill fishing is concentrated in the Antarctic part of the Atlantic. The total biomass of krill in the AA is estimated at 60.3 million tons, the value of the total allowable catch of krill is set at 5.61 million tons with a temporary catch limit of 620 thousand tons. The Commission of the participating Countries of the Convention on the Conservation of Antarctic Marine Living Resources (CCAMLR) deals with the regulation of extraction and protection of aquatic biological resources of the AA. Domestic specialists participate in international monitoring of fishing and international surveys to assess Antarctic krill stocks. Fishing statistics from available international sources are analyzed. Annually, Russia can extract about 300 thousand tons of krill in AA.

The concept of the "blue economy" goes beyond the vision of the world ocean economy solely as a mechanism for using ocean resources for economic growth, it also focuses on the sustainability of the ocean necessary for economic development. The important objectives of the Blue Economy Concept are, firstly, to understand and better manage many aspects of ocean stability, from sustainable fisheries to ecosystem health and pollution prevention. Secondly, this Concept makes society realize that the sustainable management of ocean resources requires constant cooperation between countries within the framework of various partnerships, and on a scale that has not been achieved before. Therefore, oceanic fishing, as one of the main links of the world fisheries, within the framework of the application of the "blue economy" Concept, should rely on modern components of fisheries science—forecasts of the state of the raw material base, features of the distribution and behavior of fishing objects, monitoring of the state of oceanic ecosystems in real time, taking into account the requirements of international conventions and bilateral treaties.

10.3 Conclusion

Oceanic fishing is one of the main links of the world fisheries, contributing to the systematic and sustainable development of economic sectors related to the use of the resources of the World Ocean and requiring constant comprehensive scientific support. Therefore, modern components of fisheries science, such as forecasts of the state of the raw material base, features of the distribution and behavior of fishing objects, monitoring of the state of ocean ecosystems in real time, should be taken into account when applying the Concept of the "blue economy".

The main task for our country in the research and extraction of aquatic biological resources of the Atlantic Ocean in modern conditions is to protect the interests of optimal Russian fishing at the international level. This is possible by evaluating aquatic biological resources on a strict scientific basis and developing recommendations for their rational use, taking into account the Concept of the "blue economy", the requirements of international conventions and bilateral treaties.

In total, in the oceanic regions of the Atlantic, judging by the available scientific data and fishing statistics, the domestic fleet can annually extract about 1.0 million tons of ABR without damage to the oceanic ecosystem. In addition, it should be borne in mind that there is a promising reserve—possible fishing of Antarctic krill up to 5.5 million tons annually.

References

Bandurin KV, Arkhipov AG (2017) The current state of aquatic biological resources in the oceanic research areas of AtlantNIRO. Scientific notes of RSMU 48:148–166

Chelyshev A (2021) Green color for blue economy. Retrieved from https://www.kp.ru/daily/27287/4425254/

FAO (2016) Yearbook: fisheries and aquaculture statistics (2014). Rome, Italy: FAO, 77

FAO (2020) Yearbook: fisheries and aquaculture statistics (2018). Rome, Italy: FAO, 82

Fishery economy. (n.d.). Retrieved from http://fish.gov.ru/otraslevaya-deyatelnost/ekonomika-otrasli/statistika-i-analitika

Sauskan VI, Arkhipov AG, Osadchy VM (2018) On the current state and prospects of development of the Russian fishery of aquatic biological resources in the Atlantic Ocean and the South Pacific Ocean. Fisheries 5:7–11

Sauskan VI, Arkhipov AG, Osadchy VM (2020a) Modern problems of sustainable development of the fisheries sector of the Russian economy and ways to solve them. Fisheries 6:67–72

Sauskan VI, Osadchy VM, Arkhipov AG (2020b) On the role and tasks of the national fisheries science in the development of oceanic fisheries. Fisheries 2:18–23

United Nations General Assembly (2015) Transforming our world: The 2030 Agenda for Sustainable Development. Resolution adopted by the General Assembly on September 25, 2015. NY, USA. 44 p

Why the EU supports bioeconomy research and innovation. (2020). Retrieved from https://ec.europa.eu/info/research-and-innovation/research-area/environment/bioeconomy/blue-bioeconomic_en

Chapter 11
Problems of Arrangement on the Private Land Holdings of Ponds, That Have a Hydraulic Connection with Other Water Objects

Nina Bugakova, Anna Machanek, and Natalia Yurasyuk

Abstract This article focused on the problems of arranging of ponds on the private landownership, that have a connection with natural watercourses—rivers, streams. In the process of investigating the influence of such ponds on the environment, including water resources and aquatic biological resources, is being studied. The main attention is paid to the peculiarities of the legal regulation of the construction on the territory of private land of ponds, that have a hydraulic connection with other water objects, the problems of carrying out the necessary approvals and studies aimed at observing environmental legislation. It is emphasized that in Russia there is no proper control and oversight by the state authorities over equipment in private land holdings of ponds associated with natural watercourses, which is due as to the reduction in inspections of compliance with environmental legislation during the spread of coronavirus infection (COVID-19), and due to the fact that a huge number of small reservoirs and watercourses in Russia significantly complicates the organization of state control over their condition. When organizing volunteer activities in Russia, it is necessary to take into account international experience to revitalize of volunteer movements to support environmental volunteer movements from the state, organize research activities of volunteers aimed at studying the state of small reservoirs and aquatic biological resources, as well as their accounting and registration.

Keywords Ponds on natural watercourses · Legal regulation of the construction of ponds on watercourses · Environmental volunteer movements

N. Bugakova (✉) · A. Machanek · N. Yurasyuk
Kalinigrad State Technical University, Kaliningrad, Russia
e-mail: bugakova@klgtu.ru

A. G. Arkhipov (ed.), *Sustainable Fisheries and Aquaculture: Challenges and Prospects for the Blue Bioeconomy*, Environmental Science and Engineering, https://doi.org/10.1007/978-3-031-08284-9_11

11.1 Introduction

In recent years, the practice of arranging ponds on the territory of private land holdings has been spreading in Russia. Such ponds are built for decorative purposes, for irrigation, caring for farm animals, fishing. The self-isolation regime introduced in connection with the spread of coronavirus infection had a significant impact. Citizens forced to stay on the territory of private land holdings began to engage in their development and improvement, including equipping ponds associated with natural watercourses. The impact of such ponds on the environment should not be underestimated. Ponds equipped on natural watercourses can significantly affect water resources, since a considerable amount of water evaporates from the surface of ponds. The owners of the lots of lands on which the ponds are located intensively use water from them for irrigation, what leads to the shallowing of the reservoirs the ponds are connected with. The construction of ponds is often accompanied by the construction of retaining structures what negatively affects the state of aquatic biological resources, hindering the migration of fish. Besides, the ponds can be used for poultry production. Because of this, a significant amount of organic and pathogenic microflora matter enters the watercourses. As a result, of the active access of an organic nitrogen in reservoirs a sharp increase in the biomass of phytoplankton occurs due to the rapid reproduction of blue-green algae This process is accompanied by an increase in oxygen deficiency, the development of anaerobic processes in the reservoir and mess death of aquatic biological resources (Strelkov et al. 2014). Fish farming in shallow water bodies, which include small ponds, can be accompanied by the growth of fish parasitic fauna (Avdeeva et al. 2017), especially in the absence of proper control by the owners of such water bodies. Besides, various chemicals can enter natural reservoirs through a pond as a result of excessive use of fertilizers pesticides and herbicides. For example, in Germany the problem of pesticide pollution of streams is acute (Liess et al. 2021). In that way, the wide distribution of ponds associated with natural watercourses on the territory of private land holdings can have a negative impact on the state of water objects and aquatic biological resources. At the same time, countering illegal equipping of ponds is difficult due to the lack of a clear legal status of a number of water objects that have signs of ponds, but are not related to them.

The problems of illegal equipping of ponds on natural watercourses are typical not only for Russia. In the United States, illegal equipping of ponds associated with a watercourse system in the Arkansas River is accompanied by the shallowing of this river and it is tributaries. Countering illegal equipping of ponds is to a certain extent hampered by the lack of a clear legal status of a number of water bodies that have signs of ponds, but are not related to them (Mailliard 2018).

In Russia, the issues of legal regulation of the arrangement of ponds on the territory of private land holding are not enough research. This is relevant for ponds that are associated with natural watercourses. As noted by D. O. Sivakov, due to gaps in water legislation in places where cottages and dacha communities are built, it is possible to build arbitrary dams and appropriation of parts of streams and rivers (Sivakov 2007).

It should be noted that in Russia the legislation regulating the equipment of ponds on the territory of private land holdings has undergone significant changes in recent years. Firstly, it is necessary to study the legal status of ponds on watercourses, the procedure for equipping them on the territory of private land holdings, and also to explore gaps in water and land legislation of view of ensuring proper legal protection of water objects and aquatic biological resources, which such ponds are associated with. Secondly, should be determined what organizational and legal measures can smooth out the negative trend that may arise in the conditions of uncontrolled equipment of ponds on natural watercourses. Thirdly, it is necessary to consider the of attracting volunteer environmental groups to monitor natural small water objects, connected with ponds equipped on the territory of private land holdings in order to identify facts of violation of environmental standards.

11.2 Materials and Methods

The object of the research is public relations in the field of protection of water bodies and aquatic biological resources. The subject of the research is the legal norms regulating the equipment of ponds on watercourses. Since the normative legal regulation of in that field includes many laws and by-laws that are not always consistent with each other, it is necessary to conduct a comparative analysis. It is also necessary to study the existing mechanisms of control and supervision of the equipment of ponds on watercourses. Due to the fact, that the legal norms in the field of arrangement of ponds on watercourses in Russia in recent years have changed, in the course of the study it is necessary to use historical methods.

Considering ways to counteraction illegal equipment on the territory of private land holdings of ponds associated with natural watercourses, it should be mean that control and supervision by the state in this area is limited. This is also due to the reduction in inspections of compliance with environmental legislation during the spread of coronavirus infection, and due to the fact that a huge number of small reservoirs and watercourses in Russia significantly complicates the organization of state control over their condition. We believe that it is necessary to explore the possibilities of attracting various environmental volunteer organizations to the problems of the negative impact of ponds, equipped on the territory of private land holdings, on natural reservoirs, using sociological methods. Thus, statistical studies in 2020 showed that Russian citizens consider the problem of pollution of water bodies and their banks with garbage to be very urgent (73% of respondents). However only 7% of respondents are well informed about the activities of environmental protection organizations and eco-activists (Environmental agenda: ten months before the State Duma elections. Analytical report of the all-Russian center for the study of public opinion 2020).

11.3 Results

Let's consider the main provisions of the legislation that regulate the arrangement of ponds on the territory of private landholdings connected to natural watercourses. The Water Code of the Russian Federation does not disclose the concept of "pond". In State standard 19179-73 "Hydrology of the land. Terms and definitions" there are two definitions of a pond: a pond is a shallow storage area of no more than 1 sq. m. km.; pond—a small artificial reservoir in a specially dug hole on the surface of the earth, intended for the accumulation and storage of water for various economic purposes (State committee of standards of the Council of Ministers of the USSR 1973). Ponds can be created by damming a small watercourse such as a stream, river or canal. So, in paragraph 3 of part 4 of Art. 5 of the Water Code of the Russian Federation states that the shoreline of a pond is determined by the normal retaining water level (State Duma of the Russian Federation 2006).

According to Part 2 of Art. 8 of the Water Code of the Russian Federation, ponds and watered quarries can be owned by individuals and legal entities only if they are located within the boundaries of their land plots. The question arises: can ponds associated with natural watercourses be located on the territory of private land holdings? From the point of view of water legislation, the answer to this question is negative. Ponds associated with natural watercourses can't be privately owned, since rivers, streams and other water bodies are only owned by the Russian Federation in accordance with Part 1 of Art. 8 of the Water Code of the Russian Federation. This position was generally adhered to by judicial practice until recently. In private ownership, it was possible to have ponds that have no hydraulic connection with other water bodies and are located outside the watercourses. If the pond was not isolated from other surface water bodies and had a hydraulic connection with them, in judicial practice, such ponds belonged to the property of the Russian Federation. This also applied to ponds formed on small rivers, streams and canals with the help of water-pressure structures. This practice is reflected in paragraph 23 of the Review of judicial practice of the Supreme Court of the Russian Federation No. 2 (2019), approved by the Presidium of the Supreme Court of the Russian Federation on July 17, 2019 (Presidium of the Supreme Court of the Russian Federation 2019).

However, in 2017, changes were made to the Land Code of the Russian Federation (State Duma of the Russian Federation 2017). In Art. 40 of the Land Code of the Russian Federation, a rule appeared that the owner of a land plot has the right to build ponds, including those formed by water retaining structures on watercourses, in accordance with environmental, construction, sanitary and hygienic and other special requirements. It should be emphasized that the introduction of these changes to Art. 40 of the Land Code of the Russian Federation was not accompanied by a change in the Water Code of the Russian Federation. Often, individuals do not understand that compliance with environmental requirements when equipping such ponds is a rather complicated procedure and is accompanied by numerous approvals from various government agencies. As Professor Volkov G. A. rightly notes, ponds formed by water retaining structures on watercourses arise as a result of construction (Volkov 2018). At the same time, according to tbsp. 11 of the Water Code of the Russian Federation, for the construction of a channel pond, a decision is required on the provision for use of a water body, which is a watercourse, since these works are associated with the construction of a water retaining structure, which is a hydraulic engineering, with a change in the bottom and banks of water bodies, as well as obtaining a construction permit. The procedure for coordinating construction and other works both on the watercourses themselves and in the water protection zone is regulated by various by-laws aimed at the preservation of aquatic biological resources. The person planning the production of such works shall coordinate them with the Federal Agency for Fishery, independently provide data on the assessment of the impact of the planned activity on the state of aquatic biological resources and their habitat, taking into account the fishery value of water bodies. During the approval procedure, it is necessary to assess the impact on aquatic biological resources, taking into account the calculation of the predicted damage to fish stocks and the development of measures to compensate for damage during construction. This requires the fishery characteristics of the water body, which, in turn, requires its hydrological characteristics. We do not consider these procedures to be redundant, since they are aimed at protecting natural reservoirs and aquatic biological resources. However, physical person, seeking to save money, often act unauthorized, constructing ponds on natural watercourses without designing and obtaining the necessary approvals, without conducting appropriate research. An example is the construction of ponds for breeding poultry on the Grachevka River in the village of Lyublino, Kaliningrad Region (Morozov and Werner 2018). This pond, which was built with the help of capital reinforced concrete structures, prevented the migration of fish. This activity is illegal. Thus, the Mineralovodsk City Court of the Stavropol region on May 8, 2018 in case No. 2-128/2018 decided to illegally equip a pond on a watercourse on the territory of the peasant farm of V. F. Melstein (Mineralovodsk city court. Stavropol region 2018). At the same time, V. F. Melshtein not only took water from the stream with which the pond was connected, but also illegally equipped the stream with a dam and a spillway structure. The specified work on the stream was carried out without any approvals from government agencies.

It should be noted that the use of ponds equipped on watercourses is also widely used for aquaculture purposes. From 2020, for aquaculture purposes, it is allowed to use agricultural land which occupied by watered quarries and ponds, including ponds formed by water retaining structures on watercourses, in order to carry out pond aquaculture. It should be borne in mind that the use of ponds equipped on the territory of private land holdings for aquaculture purposes in Russia is permissible only for agricultural land. It should also be noted that only individual entrepreneurs and legal entities can engage in this type of activity. In contrast to private land holdings owned by physical persons, the activities of economic entities engaged in aquaculture are controlled by state bodies. However, in relation to aquaculture, there are problems of pollution of water bodies as a result of the activities of aquaculture farms. The issue of compulsory environmental monitoring of fish farms under the agreement for the use of fish farms has not yet been resolved.

We believe that the norms of Art. 40 of the Land Code of the Russian Federation, authorizing the arrangement of ponds on private land holdings, should be specified, providing for duties of land owners in equipping ponds, formed by water retaining structures on watercourses:

- obtaining permission from state executive authorities for construction and earthworks;
- mandatory research of the hydrological and fishery state of the natural watercourse, with which the pond will be connected;
- compulsory development of a project for the construction of a pond and its coordination with state bodies exercising supervision in the field of water use, environmental and environmental legislation, as well as the protection of aquatic biological resources;
- the possibility of using ponds formed by water retaining structures on watercourses for aquaculture purposes only on agricultural land.

In addition, the Land Code of the Russian Federation should indicate that the unauthorized construction of ponds formed by water retaining structures on watercourses is prohibited. Similar norms should be included in the Water Code of the Russian Federation. In addition, at the level of the regions of the Russian Federation, it is necessary to include in the rule of land use and development, the norms for the arrangement of ponds formed by retaining structures on watercourses.

As we indicated above, in Russia it is rather difficult to organize control over the proper observance of environmental legislation when equipping private land holdings of ponds associated with natural watercourses. To identify the facts of illegal arrangement of such a pond, it is possible to use various volunteer movements. However, at the present stage, unfortunately, state bodies are not ready to cooperate with volunteers, often perceiving them as a burden with an uncertain legal status. In addition, environmental volunteering is not popular enough. A survey of 255 students of the Kaliningrad State Technical University showed that only 34% of the respondents would choose environmental volunteering from other types of volunteering. 18% of the respondents were informed about environmental volunteer organizations existing in the Kaliningrad region. The majority are not ready for the material and technical difficulties associated with environmental volunteering—77% of the respondents. 84% of the respondents do not have a sufficient understanding of the methods of recording illegal activities that damage natural reservoirs and aquatic biological resources. 56% are not ready for conflicts with the owners of ponds associated with natural watercourses. 80% have no idea what state bodies should contact in case of revealing facts of violation of environmental legislation. Thus, it is obvious that it is necessary to develop measures aimed at activating environmental volunteer movements, including in the field of checking the state of natural watercourses, with which ponds on the territory of private land holdings may be connected. This requires, firstly, a clearer legal regulation of volunteer activities. So, for example, in Russia the issue of including volunteer activity in the work experience has not been resolved, the advantages that volunteers could have in employment have not been established. Secondly, in separate regions appropriate programs should be developed, providing for the organizational and material support of volunteers. Thus, the survey of natural watercourses requires the availability of appropriate clothing and footwear, camping equipment, first aid equipment, etc. Secondly, an important area is the participation of volunteers in research activities, which may include the study of water and aquatic biological resources in streams, rivers and canals, which are associated with ponds equipped on the territory of private land holdings. Thirdly another area is the participation of volunteers in the formation of inventory of water resources, including the registration of small streams and rivers, as well as associated water bodies on the territory of private landholdings. Thus, 94% of the respondents indicated that scientific activity during the examination of natural objects helps to attract young people to participate in volunteer movements.

11.4 Discussion

It should be noted, that in science insufficient attention is paid to the problems of legal regulation of the construction of ponds formed by water retaining structures on watercourses. This to be due to a limited empirical base. The norms allowing the equipment of such ponds on the territory of private land holdings were introduced into the Land Code of the Russian Federation in mid-2017. However, it is quite difficult to identify cases of illegal construction of such ponds, because since 2020, inspections of compliance with environmental and environmental legislation by various state bodies have been limited in Russia. Accordingly, the number of court and administrative decisions in this area impedes the conduct of scientific research.

The positions of scientists regarding the possibility of equipping ponds associated with natural watercourses differ significantly. So, Sivakov D. O. points out that the taking of such ponds to private ownership can have a positive impact, contribute to the cleaning of ponds and the restoration of their ecosystem (Sivakov 2007). Volkov G. A. believes that the equipment of such ponds, as well as their transfer to private property, is unacceptable (Volkov 2018). However, pointing out the shortcomings of the legal regulation of the arrangement of ponds on watercourses, the researchers practically do not propose specific changes in the current legislation that would be aimed at unconditional compliance by the owners of ponds with environmental, construction, sanitary and hygienic and other mandatory requirements.

11.5 Conclusion

Changes to Art. 40 of the Land Code of the Russian Federation, according to which equipment is permissible in private land holdings of ponds, including those formed by water retaining structures on watercourses, do not comply with the norms of the Water Code of the Russian Federation. The indications of the legislation that the equipment of such ponds is carried out in accordance with environmental, construction, sanitary and hygienic and other special requirements on natural watercourses, are not sufficient for private persons, who often ignore the procedures for obtaining appropriate decisions and approvals, which require significant material costs from them.

We believe that the norms of the Land Code of the Russian Federation, permitting the arrangement of ponds on the territory of private land holdings, should be specified, providing for a number of obligations of land owners when equipping ponds formed by water retaining structures on watercourses and indicate that the unauthorized construction of ponds formed by water retaining structures on watercourses is prohibited. Similar norms should be included in the Water Code of the Russian Federation. In addition, at the level of the regions of the Russian Federation, it is necessary to include in the rule of land use and development, the norms for the arrangement of ponds formed by retaining structures on watercourses.

We believe that in this area it is necessary both to strengthen state control and supervision, and to widely involve environmental volunteer organizations to monitor small water bodies, as well as to identify the facts of illegal equipment of ponds associated with natural watercourses. To intensify volunteering in Russia, international experience should be taken into account. It is necessary to support environmental volunteer movements from the state, organize research activities of volunteers aimed at studying the state of small reservoirs and aquatic biological resources, as well as their accounting and registration.

In the context of limited state control and supervision associated with the spread of coronavirus infection, prolonged neglect of equipment problems in private landownership of ponds associated with natural watercourses can lead to irreparable consequences for natural water bodies.

References

Avdeeva EV, Evdokimova EB, Zaostrovtseva SK (2017) Current state of studying fish parasitofauna of the Kaliningrad region water bodies. News Kaliningr State Tech Univ 4:24–61 (in Russian)

Environmental agenda: ten months before the State Duma elections. Analytical report of the all-Russian center for the study of public opinion (2020). https://wciom.ru/analytical-reports/analiticheskii-doklad/ehkologicheskaja-povestka-zadesjat-mesjacev-do-vyborov-v-gosdumu (in Russian)

Liess M, Liebmann L, Vormeier P, Weisner O, Altenburger R, Borchardt D, Brack W, Reemtsma T (2021) Pesticides are the dominant stressors for vulnerable insects in lowland streams. Water Res 201:1–11 (in Russian)

Mailliard K (2018) Illegal ponds in Colorado: more than just a drop in the bucket? University of Denver Water Law Review, November 19. https://duwaterlawreview.com/illegal-ponds-in-colorado-more-than-just-a-drop-in-thebucket (in Russian)

Mineralovodsk city court. Stavropol region (2018). Decision of the Mineralovodsk city court (in Russian)

Morozov K, Werner M (2018) In the village of Lyublino, fish may die due to illegal construction. Kaliningrad News. http://vesti-kaliningrad.ru/vposelke-lyublino-iz-za-nelegalnoj-postrojki-mozhet-pogibnut-ryba/ of the Stavropol region No. 2-362/2018. Mineralovodsk city court. Stavropol region. 2-362, p 3 (in Russian)
Presidium of the Supreme Court of the Russian Federation (2019) Review of judicial practice of the Supreme Court of the Russian Federation. 2 (in Russian)
Sivakov DO (2007) Water law: educational and practical guide. Yustitsinform, Moscow, Russia (in Russian)
State Committee of Standards of the Council of Ministers of the USSR (1973) State standard of the Union of SSR 19179-73 "Hydrology of land. Terms and definitions" (October 9, 1973 No. 2394). USSR State Committee for Standards, Moscow, Russia (in Russian)
State Duma of the Russian Federation (2006) Water code of the Russian Federation (June 3, 2006 No 74). Russian newspaper, Moscow, Russia (in Russian)
State Duma of the Russian Federation (2017) Land code of the Russian Federation (October 25, 2001 No 74). Russian newspaper, Moscow, Russia (in Russian)
Strelkov KE, Lushkin IA, Filenkov VM (2014) Causes and consequences of flowering water sources used for drinking water supply. Bull Nizhniy Novgorod State Eng Econ Inst 12(43):79–84 (in Russian)
Volkov GA (2018) Ponds and watered quarry: Who is the owner? Further discussion. Saratov State Law Acad Bull 3(122):173–179 (in Russian)

Chapter 12
Problems of Implementation of the Program for the Development of the Fisheries Industry of the Russian Federation

Victoriya V. Dorofeeva, Elena V. Klippenshtein, and Yulia S. Morozova

Abstract Problems, appeared within intermediate stages of the implementation of the state program named "Development of the fisheries industry", approved on April 15, 2014 by the decree of the government of the Russian Federation have been overviewed in this article. Necessity of the realization of analyzed program is due to the high rate of reduction of the fishing fleet, rising lack of production power, low grade of processing of catches as well as lack of innovations within fisheries industry. The state program "Development of the fisheries industry" is designed to insure a departure from raw-material export direction by stimulation the production with a high share of added cost through the development of scientific and technical potential of the fisheries industry. Implementation of the program led to a sharp increase of investments in modernization of current fishing fleet and construction of new vessels due to the change in the principle of allocation of quotas. It contributes to negative financial consequences for investment companies that can't implement fishing and fulfil their financial obligations to financial institutions in a timely manner due to delayed commissioning of fishing vessels. Changes in principles of quotas allocation and adding new mechanism named "keel quotas" deprive fishing enterprises of a certain part of the potential incomes, with support of which these enterprises intended to cover previous investments in modernization, re-equipment of the fleet and onshore processing. It is necessary to note that in present times this program supports only shipbuilding complex, ignoring financial interests of both fishing and processing enterprises. Position of the author is based on necessity of implementing not only modernization and re-equipment of the fleet, but also develop technologies of deep and waste-free processing of extracted aquatic biological resources in terms of current subprograms.

V. V. Dorofeeva · E. V. Klippenshtein (✉)
Kaliningrad State Technical University, Avenue Sovetsky, 1, 236022 Kaliningrad, Russia
e-mail: evklipp@mail.ru

Y. S. Morozova
Kamchatka State Technical University, St. Kluchevskay, 35, 683003 Petropavlovsk-Kamchatsky, Russia

A. G. Arkhipov (ed.), *Sustainable Fisheries and Aquaculture: Challenges and Prospects for the Blue Bioeconomy*, Environmental Science and Engineering, https://doi.org/10.1007/978-3-031-08284-9_12

Keywords Aquatic biological resources · Fisheries industry development program · Industrial fishing · Keel quotas

12.1 Introduction

The state program of the Russian Federation "Development of the fisheries industry" adopted in 2014 was supposed to ensure balanced development of whole complex: from shipbuilding to realization of fishing products. But analysis of its implementation allows to conclude that real shifts in renewal of fishing power, as suggested by the most significant subprogram, have begun only after 2018, when other rules for fixing catch quotas were. However, experts point to unwillingness of Russian shipbuilding enterprises to fully and on time fulfil the concluded contacts for construction of new ships. Current situation negatively effect on financial state of fishing companies, which could not catch the planned volumes of aquatic biological resources.

It is necessary to fix program in terms of prioritization: in addition to updating and modernization of fleet as main subprogram, efforts should be focused on innovations that will provide depth and non-waste processing of extracted aquatic biological resources.

12.2 Materials and Methods

The article has attempted to intermediate results of the implementation of the state program of the Russian Federation "Development of the fisheries industry" in order to determine the achievement of target indicators and, if necessary, develop of recommendations for adjusting the goals and mechanisms of certain subprograms. During writing the article, the authors used expert assessments, published in different sources, selected though content analysis. When forming a sample of sources for the most objective presentation of the opinions of specialists about dynamics of the development of fisheries industry in terms of program implementation, publications in the media were considered, representing the interests of fishery producers of all fishing regions in the Russian Federation, as well as regulatory documents of the Federal Agency for fisheries.

12.3 Results

For a long period of time, the Russian Federation has been a leader in extraction and export of aquatic biological resources and one of the five countries with the highest volumes of industrial fishing (FAO 2021). According to the Federal Agency for fishery in the Russian Federation, in 2019 the total volume of fishing for aquatic biological resources amounted to about 5 million tons. Volume of extraction of aquatic resources in 2019 was only 2.2% less than in 2018 (5.03 million tons). At the same time, in 2018 there was a maximum volume of extraction for last 26 years (Scopintseva 2019).

With a relatively high rate of operating profitability (over 30%) and high catch rates, enterprises are in no hurry to invest in development of production. Despite good financial performance and government efforts, the structure of fisheries production remains unchanged. The yield of products from aquatic biological resources remains at about 65%, which indicates a low degree of processing of catches on ships and lack of innovations. About 90% of food products account for the production of frozen cut and uncut products (Catch volume 2019).

To form stable and balanced system "extraction-processing-consumption of fish products" back in 2014, the state program of the Russian Federation of April 15, 2014 № 314 named "Development of fisheries industry" was adopted, which assumed the extension of implementation of scientific researches, development of scientific and technical potential of fisheries complex and avoiding the raw-material export direction by stimulation of production with high share of added value and creating favorable conditions for doing business and attracting investments in the industry (Government of the Russian Federation 2014). However, analysis of indicators of program realization and opinions of expert society allows to state, that problems of the industry are still unsolved.

In 2018, the Federal Agency for fisheries conducted a campaign to re-fix the shares of quotas for extraction aquatic biological resources for next 15 years, as a result 3939 agreements were negotiated (Federal Agency for fisheries 2019). In addition, two main stages of the campaign to secure the shares of investment quotas have been implemented. Within the framework of the implemented state program, the construction of 43 large and medium tonnage ships of fisheries fleet has been approved. In the near future, until 2023, it is planned to build 33 ships, the part of which are under construction process, and by 2025, it is planned to build 11 more vessels at domestic shipbuilding enterprises. Of the 43 projects mentioned, 18 ships are planned to be built for the Far East fishery basin, and 25 ships for the Northern basin. The volume of investments in shipbuilding industry is 166 billion rubbles (Alekseeva et al. 2020).

These measures of state regulation of fisheries complex were supposed to solve issue of moral and physical deterioration of material and technical, but at the present time there is a consistent reduction in production capacity of the fleet in the main basins.

Fishing fleet is the basis for material and technical base of the industry. It accounts more than 70% of fixed assets and 90% of fish extraction. The ships produce 92% of total volume of frozen products, 96% of fishmeal and 15% of canned food. The main part of fleet (71%) is concentrated in the Far East, 16% in the Northern basin, 8% in the Western basin, 3% in the Azov-Black Sea basin, 2% in the Volga-Caspian basin (Experts stated 2020).

From 2000 to 2018, the composition of Russian fishery fleet in quantitative terms has decreased by 25% (How to build 2020). According to different assessments, in case if current tendencies towards decommissioning of ships continue, in the near future it will be an acute shortage of production capacity that will lead to huge limitation in increasing of volumes of aquatic biological resources extraction. And this is despite the fact that in terms of using the investment quotas, enterprises have already concluded contracts for fishery shipbuilding, taking into account the fact that it will take at least 5 years to build the ship (Experts stated 2020).

The "keel quota" program is implemented by United Shipbuilding Corporation (USC)—the largest shipbuilding company in Russia. The corporation includes about 40 industry enterprises: the mail shipbuilding and fixing yards and leading project and construction bureaus (USC 2020). At the present time, most of the domestic shipbuilding complex is consolidated in the basis of USC.

The main place of concentration of investment program of the Federal Agency for fishery and the Ministry of Industry and Trade are shipyards of St. Petersburg and the Leningrad region. St. Petersburg's shipyards will build 20 modern vessels for fishery fleet of Russia (Ships of St. Petersburg 2020).

In total, eleven shipyards and shipbuilding factories are in "keel quota" program at present time. At different stages of building (some have already been handed over to customers, some wait for quota) there are 67 units of the fishing fleet (Volovik 2020).

Meanwhile, most of fishing vessels which are building at Russian shipyards under the program of investment quotas, will be delivered to customers with delay. Building process of 27 fishing vessels, the laying of which took place in 2017–2018, is significantly behind schedule. The laying of 16 more fishing vessels, expected in 2021, will definitely violate the approved planes. And this is despite the fact that fishing enterprises, in accordance with their contractual obligations, strictly follow the construction financing schedules (Systemic "long-term construction 2020).

Thus, violation of the terms of construction of fishing vessels occurs due to circumstances beyond the control of investors. According to the information of shipyards representatives, for all vessels under construction, the backlog of the schedule will be from six month to two years (Possibility of termination 2020).

Shipbuilders point to three main reasons for delays in construction. Firstly, it is the financial conditions of shipyards: they often have to "cover" the costs of building some ships with advance payments for other payments. Secondly, there is a lack of staff. A significant role was played by the fact that initially all work was compressed in a very narrow period of time and, in fact, concentrated a large number of vessels on a limited number of shipyards. Thirdly, due to the pandemic, shipyards faced delays in supplies of import components, and in some cases with a refusal to cooperate on previously concluded contracts with foreign subcontracts. Because of the closure of international traffic, deadlines for the execution of technological operations requiring the participation of foreign specialists (sea trials, commissioning, etc.) are disrupted.

Fishery enterprises strictly follow the construction financing schedule. Interest on borrowed loans and servicing bank guarantees for the period from 2017 to 2019 is at least 10–12 billion rubles and the ships have not been built yet (Systemic "long-term construction 2020). It is necessary to point out the position of the Ministry of Industry and Trade, representatives of which in 2015–2016 were proving to the government the readiness of the industry under their jurisdiction to build dozens of fishing vessels, but so far only 2 have been built (Possibility of termination 2020).

Therefore, investors suffer huge losses because of maintenance of the bank guarantees and providing of ships construction. Moreover, these enterprises will incur additional losses, since they will not receive quotas on time.

It is necessary to note that representatives of fishery industry, since 2011, have actively opposed the implementation of "keel quota" investment program, referring to both foreign and domestic experience (Ponyavin 2019). Large-scale pumping of the industry with capital investments lead to overcapitalization of fishery complex, that can cause a cap between the fishing capacity of the fishing fleet and extraction of aquatic and biological resources. And inertia of investments into fishing fleet will only widen this gap. It was also pointed out that a change in historical principle of quotas distribution and introduction of a new mechanism of the quotas distribution can deprive the fishing industry of certain part of potential income, with the help of which these enterprises planned to cover previous investments into modernization and re-equipment of the fleet and coastal processing. In addition, financing of construction of new vessels and servicing of loans occurs at the expense of the proceeds from the remaining quota, that negatively effect on financial conditions of those fishing enterprises, which actively took part in implementation of the program.

A number of representatives of the fishing industry point to the fact, that the allocated investment quotas are enough to load a maximum 50% of capacity of new ships and factories, which are being built under analyzed program (Ministry of agriculture 2020).

Against the background of this situation, a number of shipbuilding enterprises that took part in the program and found themselves in a difficult financial situation require a number of government decisions to further develop the practice of "keel quotas" in terms of increasing the share of investment quotas from 20 to 50% for fish and 100% for crab, as well as expand this practice to catch shrimp, trumpeters, sea urchins and other valuable species of biological resources, and at the same time prohibit the operation of fishing vessels over 30 years old in the EEZ of the Russian Federation. The proposed measures were supported by a number of relevant ministries, but were extremely negatively perceived by representatives of the fishing industry and regional authorities, where the fishery complex plays a significant role in the regional economy (Burlakova 2020).

So, in particular, the relevant departments of the Kamchatka Territory indicate that as positive effects from the implementation of the investment quotas program, one can note an increase in the investment activity of enterprises participating in the program, an increase in the predicted tax return on projects being implemented and the creation of modern work places. At the same time, for a comprehensive assessment of the effect obtained from the implementation of the program of investment quotas for the development of fisheries, it should be noted that the allocation of 20% of quotas for investment purposes has already led to a decrease in the resource base of fishing enterprises, and the proposed increase in the share of the total allowable catch of crabs provided for investment purposes from 50 to 100%, and aquatic biological resources from 20 to 50% will lead to its significant reduction, as well as significantly and negatively affect the implementation of investment projects (Open Letter 2020). A reduction in the catch of aquatic biological resources will also entail a decrease in payments to the budgets of all levels due to a reduction in payments for aquatic biological resources.

In general, the proposed solution to increase the share of the total allowable catch (provided for investment purposes) of aquatic biological resources up to 50% and crabs to 100%, and due to this, reducing of quota volume for fishery enterprises, will inevitably worsen their financial condition that can lead to massive reduction of staff, as well as will not allow to enterprises participating in program to provide the payback of ongoing investment projects within planned period of time, that will entail a deterioration in the investment climate.

Speaking about reserves for the growth of extraction volumes, many experts note that among the fishing fleet projects, proposed by designing institutes of USC, there are no solutions aimed at the development of unprofitable kinds of aquatic biological resources in Russian economic area, as well as profitable fishing in the World ocean.

It is necessary to note, that the fishery industry, initially, expressed doubts about tha ability of domestic shipyards to accept orders for construction of competitive vessels (Fashutdinov 2020).

At the present time, it is difficult to conclude how realization of the investment program will affect on further prospects of fisheries. For example, trawlers which are already built and transferred to fishing collective farm named after Lenin, for a number of objective and subjective reasons can not reach production capacity to assess the efficiency of implemented investments.

In connection with the facts listed above, it can be stated that the investment program "Development of the fisheries industry" is primarily aimed at protecting the interests of domestic shipbuilders. It should be noted that during the construction of the domestic hull of a fishing vessel, investors almost 90% chose foreign filling of ship and fishing equipment, which reduces the economic effect from the implementation of the program (Alekseeva et al. 2020). At the same time, the growth potential of the fishing industry by increasing the catch of the most profitable species of the fishery (pollock, herring or salmon) is currently limited. Therefore, in our opinion, government measures to support the development of the fishery complex should be concentrated not only on the renewal and modernization of the fleet, but also on innovations that will ensure the depth and waste-free processing of extracted aquatic biological resources.

12.4 Discussion

During analysis of expert opinions and discussing of the issue of processing aquatic biological resources, we have detected a clear underestimation of variety of fishing methods, taste and technological characteristics of raw materials, and, consequently, products, uneven distribution, remoteness from places of consumption (Ponyavin 2019; Scopintseva 2019; Experts stated 2020). So it is problematic to set general priorities and principles in determination of processing depth. In this case, we suppose that it is necessary to solve this problem simultaneously with organization of the processing of waste, generated during extraction of aquatic biological resources.

According to various assessments, almost 1.7 million of 5 million tons of extracted aquatic biological resources go "overboard" in the form of waste (Mikhaylov 2020; Makoedov and Kozhemyako 2020; Experts stated ... 2020). According to industry experts, there are technologies for waste-free fish processing in Russia, but their use is hampered by a number of restrictions. Firstly, fishermen throw away not only production waste, but also valuable fish species for which a company or a particular vessel does not have quotas. Secondly, coastal fishing enterprises are deprived of the opportunity to legally release and deliver products from fish waste to the shore. In this case, the fishermen are fined, since according to the federal fisheries law, the products of coastal fishery must be delivered to the shore in live, fresh or chilled form. Thirdly, the production of products from the waste of processing aquatic biological resources has a low profitability, which makes this type of activity unattractive for the fishery.

So, we detect the following directions of further development of the state program "Development of the fisheries industry" for updating the material and technical base of the fishing complex:

- expanding the provision of investment quotas for the construction of low-tonnage vessels;
- using the mechanisms of investment quotas not only in shipbuilding and the development of coastal processing, but also for the development of waste-free production and increasing the depth of processing of aquatic biological resources;
- introduction of subsidizing mechanisms for the production of products from waste generated during the extraction of aquatic biological resources;
- changing the organization of processing of waste which is generated during the extraction of aquatic biological resources.

12.5 Conclusion

Due to non-fulfillment of the indicators of the subprogram for the modernization of the fishing fleet, a decrease in the catch of aquatic biological resources is observed, which entails a deterioration in the financial condition of the enterprises of the fishing industry and a decrease in payments to the budgets of all levels due to a reduction in payments for aquatic biological resources.

There is an opinion that investment program "Development of the fisheries industry" is primarily aimed at protection of interests of domestic shipbuilders. But even in such favorable competitive conditions, vessels under construction are almost 90% filled with ship and fishing equipment due to imports, which reduces the economic effect from the implementation of the program. The projects of fishing vessels proposed by the USC design institutes are not focused on the development of unprofitable types of aquatic biological resources in the Russian economic zone and profitable fishing in the World Ocean. At the same time, all experts emphasize that the growth potential of the fishing industry by increasing the catch of the most profitable species of the fishery (pollock, herring or salmon) is currently limited.

Therefore, in our opinion, government measures to support the development of the fishery complex should be concentrated not only on the renewal and modernization of the fleet, but also on innovations that will ensure the depth and waste-free processing of the extracted aquatic biological resources, together with the organization of processing waste, which is generated during the extraction of aquatic biological resources.

References

"Eastern Shipyard" (Vladivostok) laid the first crabber under the program, quotas in exchange for investments" (n.d.) (2020). https://portnews.ru/news/287551/ (in Russian)

«Pella» continues the construction of industry vessels (n.d.) (2020). https://otradnoevsz.ru/news?id=793 (in Russian)

Alekseeva OE, Babchuk EV, Krulyov AA (2020) Analysis of the implementation of the state investment program and the prospects for the construction of fishing vessels at domestic shipbuilding enterprises. Works Krylovskiy State Sci Center 1(391):219–230 (in Russian)

An open letter from fishermen to President Vladimir Putin (n.d.) (2018). https://www.mk.ru/politics/2018/07/02/otkrytoe-pismo-rybopromyshlennikov-prezidentu-vladimiru-putinu.html (in Russian)

At «Nobel Brothers' shipyard» the trawler "Vympel" was launched (n.d.) (2021). https://sudostroenie.info/novosti/28671.html

Burlakova E (2020) Gleb Frank's company proposed the fish industry reform to government. https://www.vedomosti.ru/business/articles/2020/08/04/835961-kompaniya-gleba-franka-predlozhila-pravitelstvu-reformu-ribnoi-otrasli (in Russian)

Experts announced a possible acute shortage of fishing vessels in the Russian Federation in 2020 (2020). https://www.interfax.ru/russia/688353 (in Russian)

FAO (2021) The state of world fisheries and aquaculture–2020. Measures for increasing the stability. https://doi.org/10.4060/ca9229ru

Faskhutdinov (n.d.) (2020) New vessels will never be paid off… https://www.korabel.ru/news/comments/novye_suda_ne_okupyatsya_nikogda.html (in Russian)

Federal Agency for Fishery (n.d.) (2019). http://fish.gov.ru/obiedinennaya-press-sluzhba/obzor-smi/19-press-tsentr?start=1580 (in Russian)

Fishing catch volume in the Russian Federation in 2019 (n.d.) (2020). https://www.moshol14.ru/press-centr/novosti-rynka/dobycha-ryby (in Russian)

Government of the Russian Federation (2014) State program of the Russian Federation of April 15, 2014 г. № 314 "Development of fisheries industry" (with changes, of April, 2020). http://docs.cntd.ru/document/499091766 (in Russian)

How to build new fishery fleet (2020) https://plus.rbc.ru/news/5f6478a87a8aa9450390f8b7 (in Russian)

Makoedov AN, Kozhemyako ON (2020) Problems of the fishery complex of the country and the main tasks to solve them. http://arktikfish.com/index.php/stati-po-akvakulture/22-problemy-rybokhozyajstvennogo-kompleksa-strany-i-osnovnye-zadachi-po-ikh-preodoleniyu (in Russian)

Mikhaylov A (2020) Unprofitable catch. https://rg.ru/2020/06/23/reg-szfo/pochemu-tret-dobytoj-rybakami-ryby-vybrasyvaetsia-v-more.html (in Russian)
Ministry of Agriculture against the reform of fishing industry, proposed by RPPK (2020). https://m.ru.investing.com/news/economy/article-1992595 (in Russian)
Ponyavin D (2019) Historical principle is world's experience. https://fishnews.ru/rubric/investkvoty/11399 (in Russian)
Scopintseva E (2019) Fishmen are not looking for added value. Econ Life 24. https://www.eg-online.ru/article/401610/ (in Russian)
Shipbuilders of St. Petersburg will hand over 20 modern fishing vessels (n.d.) (2020). https://sudostroenie.info/novosti/32338.html (in Russian)
Some customers are considering the possibility of terminating contracts for the construction of fishing vessels in the Russian Federation (n.d.) (2020). https://portnews.ru/news/296988/ (in Russian)
Systemic "long-term construction' increases the financial burden on the fishing industry (n.d.) (2020). http://fishkamchatka.ru/articles/russia/35459/ (in Russian)
Two new crabbers are laid at Khabarovsk shipyard (n.d.) (2021). https://www.todaykhv.ru/news/economics-and-business/26279/ (in Russian)
USC (United Shipbuilding Corporation) (n.d.) (2021). https://www.aoosk.ru/companies/ (in Russian)
Volovik K (2020) What fishing vessels are being built in Russia. https://www.korabel.ru/news/comments/kakie_ryboloveckie_suda_segodnya_stroyatsya_v_rossii (in Russian)

Chapter 13
Natural and Artificial Reproduction of Sterlet in the Cheboksary Reservoir

A. E. Minin and L. M. Minina

Abstract For the first time in 40 years of the existence of the Cheboksary reservoir, a study was made of the conditions and efficiency of sterlet reproduction in this reservoir. Based on a comparison of the results of natural spawning and artificial releases of sterlet juveniles into the Cheboksary reservoir, the effectiveness of each of them in the reproduction of a valuable species of aquatic biological resources is shown. A revision of the existing natural spawning grounds for sterlet, taken from the existing Fishing Rules, was carried out in comparison with pre-damming conditions according to the Atlas of sturgeon spawning grounds. Comparison of the growth of sterlet in the Cheboksary reservoir at the present time and in the period before the formation of the reservoir showed a significant increase in its rate. The rate of linear and weight growth in the Cheboksary reservoir is at one of the highest levels compared to other reservoirs in the Upper and Middle Volga. Calculations were made of the commercial return from the downstream larvae on the Oka river according to data for the period before mass releases of sterlet juveniles. Systematized and analyzed data on the artificial reproduction of the studied fish species in the Cheboksary reservoir for the period 2009–2020. The annual commercial return from the natural reproduction of sterlet in the Cheboksary reservoir was 2.7 tons, and from artificial reproduction—6.4 tons. The effectiveness of the ongoing fish-breeding measures to restore the stock of sterlet is shown.

Keywords Sterlet · Cheboksary reservoir · Natural reproduction · Restocking

13.1 Introduction

Many large rivers are currently blocked by dams. The construction of the dams affects the river drain mode and significally changes the aquatic ecosystems (Guti and Berczik 2014; Górski et al. 2012). River systems, including their floodplains

A. E. Minin (✉) · L. M. Minina
Nizhny Novgorod Branch of Russian Federal Research Institute of Fisheries and Oceanography, Moskovskoe Shosse St. 31, 603116 Nizhny Novgorod, Russian Federation
e-mail: aeminin@mail.ru

A. G. Arkhipov (ed.), *Sustainable Fisheries and Aquaculture: Challenges and Prospects for the Blue Bioeconomy*, Environmental Science and Engineering, https://doi.org/10.1007/978-3-031-08284-9_13

and delta, are one of the most biologically diverse and productive ecosystems on the planet (WWF 2018). The construction and operation of dams leads to such a negative consequences to nature as a violation of the natural regime of the drain, fragmentation of river basins ecosystems, violation of migration pathways and a change in the habitats of the fauna and flora, transformation of the channel processes, etc. (Middelkoop et al. 2005; Torabi Haghighi et al. 2014). If the productivity of regulated rivers, as a rule, increases due to the accumulation of organic matter (Baturina 2019), then biodiversity is significantly reduced due to decline of rheophilic species in ecosystems (Zalepukhin 2009).

In the Volga reservoirs in the structure of ichthyocenoses there are changes regarding the reduction of the number or the complete disappearance of such rheophilic species as sterlet and nase (Schletterer et al. 2019). In many Volga regions, these species are listed in the Red Data Books. Sterlet, as a commercially valuable species of fish, in addition to disturbing habitat and breeding conditions, is also under high pressure from poaching (Kalmykov 2005; Bondarchuk 2016).

An analysis of the available historical data on the sterlet catches in Russia indicates its high abundance in the recent past. At the end of the 19th and the beginning of the twentieth century, sterlet catches reached 3–4 thousand tons, of which about half were from the Volga basin (Meisner 1933). The regulation of the Volga radically changed the conditions for the existence of the sterlet, worsened the conditions for its reproduction, and made serious changes in its biology. In reservoirs, sterlet, an obligate rheophilus, moved to areas where elements of the river regime have been still remained (Kuzmina 2005). Throughout the twentieth century, spawning areas of sterlet have been steadily decreasing, and the current scale of natural reproduction cannot ensure maintenance of the sterlet populations at an ecologically optimal level. Ultimately, in most of the reservoirs of the Volga cascade, in the lake-like ones in the first place, the sterlet gradually disappeared. (Boldina 1962). Afanasyev (1980) argued 40 years ago that sterlet almost completely disappeared from the ichthyofauna of the Upper Volga. But even in the unregulated Lower Volga, there is a significant decrease in sterlet stocks due to illegal fishing (Kalmykov 2005).

At present, only in the Middle Volga, sterlet is preserved in the Volga-Kama reach of the Kuibyshev reservoir in more or less significant quantities (there is commercial fishery on this specie), since there are spawning grounds yet in the channel sections of the Volga and the lower reaches of the Kama, but they are increasingly decreasing due to pollution and silting (Tairov et al. 2016; Kalayda et al. 2019; Graninet al. 2020).

On the section of the river The Volga in the area of the future Cheboksary reservoir and in the first years of its existence, sterlet was a rather numerous species and its catches in some years reached 7 tons (Shibaev et al. 1989). Due to the catastrophic decline in sterlet stocks due to silting of its spawning grounds in the Cheboksary reservoir, fishing for it was closed in 2003. Currently, the stock of this fish species is at a very low level.

Thus, the restoration and increasing the number of sterlet in the Volga reservoirs in modern conditions is possible only due to its artificial reproduction (Mamontov and Gepetskiy 2001) and the subsequent release of juveniles (Kuzmina 2005; Kalayda et al. 2019). That is why in recent years great importance has been attached to the factory breeding of sterlet (Mikhailova 2001; Bykov 2003; Bondarchuk 2016).

In the conditions of the joint presence of natural and artificial reproduction of sterlet, there is a problem of determining the effectiveness of fish-breeding measures of the artificial release of juvenile sterlet into reservoirs. One of the methods of its solution is the tagging of released fish. Trimming or perforation of fins, the introduction of radioactive isotopes are used for this. But as it turned out, they are ineffective. In 2019, Siberian scientists began experiments on tagging sturgeon juveniles with chips, but these works have not yet been tested in real natural conditions ("Siberian scientists" 2019).

In this study, an attempt was made to assess the contribution of natural and artificial spawning of sterlet of the Cheboksary reservoir based on a comparison of the abundance of fish from downstream juveniles of this species in the period before the start of mass releases (until 2009) and analysis of the survival rate according to the accepted coefficients of commercial return from released juveniles (statistical data). In addition, the paper provides information on the state of the sterlet population, its growth rates and maturation during periods before the creation of the Cheboksary reservoir and at present.

Recent studies on the ecology and breeding characteristics of the sterlet in the Cheboksary reservoir were carried out in the early years of the formation of the reservoir in the 1980s. In this regard, the present research is very relevant.

13.2 Materials and Methods

The aim of the work was to compare the effectiveness of natural fish spawning and the results of artificial reproduction of sterlet in the Cheboksary reservoir. To achieve the goal, the following tasks were set:

- characterize the Cheboksary reservoir in terms of the potential for natural spawning of sterlet;
- to assess the state of the sterlet population of the Cheboksary reservoir;
- give a brief biological characteristics of the sterlet of the Cheboksary reservoir;
- calculate the number of sterlet in the commercial return based on the results of larval surveys in the period before the start of mass releases (until 2009);
- assess the results of releases of young sterlet into the Cheboksary reservoir (2015–2020);
- compare the efficiency of natural sterlet spawning and the results of its artificial reproduction in the Cheboksary reservoir.

To compile the list and location of sterlet spawning grounds during the pre-damming period, materials from the Atlas of sturgeon fish were used (Khoroshko et al. 1971). The current number and location of spawning grounds is taken according to Appendix No. 6 to the Fishing Rules of the Volga-Caspian Fishing Basin (Ministry of Agriculture of the Russian Federation 2019).

The assessment of the sterlet stocks status was carried out according to the annual monitoring ichthyological surveys carried out by the Nizhny Novgorod branch of the Federal State Budgetary Scientific Institution VNIRO (NizhegorodNIRO) using a bottom trawl for the period 1980–2020.

The length of a caught sterlet was measured from the tip of the snout to the center of the fork on the tail (fork length), the total mass of the fish is taken as the mass. Determination of age in sterlet was carried out according to thin sections of cuts of marginal rays of pectoral fins. The rate of linear and weight growth in the 2000s (own data) was compared with the data for the pre-storage period (Afanasyev 1985). Lukin's six-point scale was used to determine the stages of maturity of the gonads in females (Lukin 1941).

Due to the fact that the concentration of sterlet differs significantly in the rivers Volga and Oka, and they represent an inextricable waterbody, the biomass index was calculated as a weighted average based on the areas of the Cheboksary reservoir (107.6 thousand ha) and the Oka river (12.5 thousand ha), according to trawl surveys.

To assess the natural spawning of sterlet, we used data from NizhegorodNIRO on surveys of downstream larvae in 2001, i.e. until the period of regular releases of juveniles into the reservoir. Statistical data on sterlet fry releases from 2009 to 2020 provided by the Fishery department.

The fishing return coefficients for calculating the number of individuals of the commercial size are taken according to Annex 2 of Methods for calculating the amount of damage caused to aquatic biological resources (Ministry of Agriculture of the Russian Federation 2020).

13.3 Results

13.3.1 Characterization of the Cheboksary Reservoir in Terms of the Potential for Natural Spawning of Sterlet, Revision of Spawning Grounds

The Cheboksary reservoir, the youngest in the Volga cascade, is located on the border of the Upper and Middle Volga. It was formed as a result of the overlap of the Volga by a hydroelectric dam in 1980 near Novocheboksarsk city (Chuvash Republic). The design assignment provided for bringing the water level to 68 m. However, due to technical and financial problems, the reservoir has existed for more than 40 years at FSL (full supply level) mark of 63 m, i.e. 5 m below the design mark. The length of the reservoir is 294 km, the area (at the 63 m) is 108 thousand hectares, the average depth is 3.5 m. The Oka River is an integral part of the Cheboksary reservoir and up to 45 km from the mouth it is included in the flooding zone.

The formation of water masses of the Cheboksary reservoir occurs mainly due to the influx of water from the Gorky reservoir and the Oka river—sources of almost equal volumes. A distinctive feature of the Cheboksary reservoir is a high water

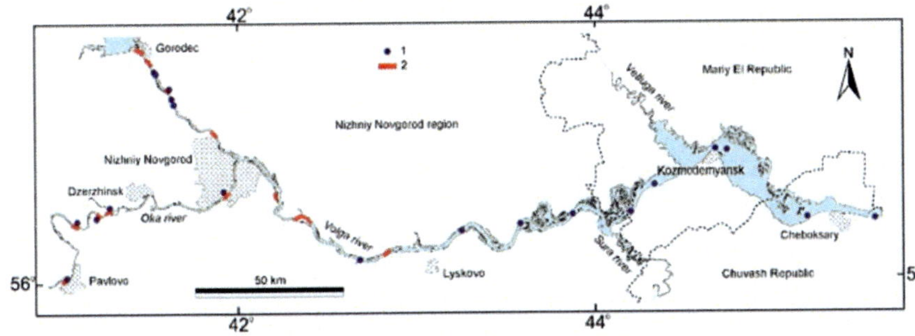

Fig. 13.1 Location of spawning grounds for sterlet in the Cheboksary reservoir: 1—spawning grounds before the creation of the reservoir; 2—modern spawning grounds. *Source* Compiled by the authors

exchange coefficient—19.8 up to 32.2. The planned water exchange provided for an average value of up to 6 times a year.

A significant decrease of the current velocity in the Volga after the creation of the Cheboksary reservoir caused siltation of spawning grounds for sterlet and other lithophilous fish species in most part of the river. Sterlet spawning grounds are sections of a river with a fast current and dense soil (gravel, sand). The number of spawning grounds after the construction of the dam was halved (see Fig. 13.1).

Sterlet spawning grounds are distinguished between channel and in-shore (Afanasyev 1980). The channel spawning grounds are the most productive, but in general it is they that have largely disappeared during the creation of the reservoir. Where the spawning grounds were partially preserved, the flow rates was sufficient for successful spawning only in a short period of flood. After regulation, the temperature regime of the Volga also changed—the heating of water in spring in reservoirs is much slower. Thus, after the creation of the Cheboksary reservoir, extremely unfavorable conditions developed for the natural reproduction of sterlet.

13.3.2 Assessment of the Cheboksary Reservoir Sterlet Population State

Earlier, in the pre-damming period (up to the 1980s), in the 1980s and up to the mid-1990s, the sterlet was a commercial specie within the boundaries of the present Cheboksary reservoir. The fishery mastered up to 5 tons of this fish species. Since the mid-1990s there was a sharp decline in catches (see Fig. 13.2).

When the process of calculating the fish biomass index was carried out, sharp jumps in quantitative indicators were found in the Volga section of the Cheboksary reservoir. The study of this phenomenon showed that when the sterlet decreases in the Volga section, the indicators increase in the Oka section of the reservoir, i.e. fish migrate from less flowing areas to faster ones. Therefore, to analyze the dynamics of

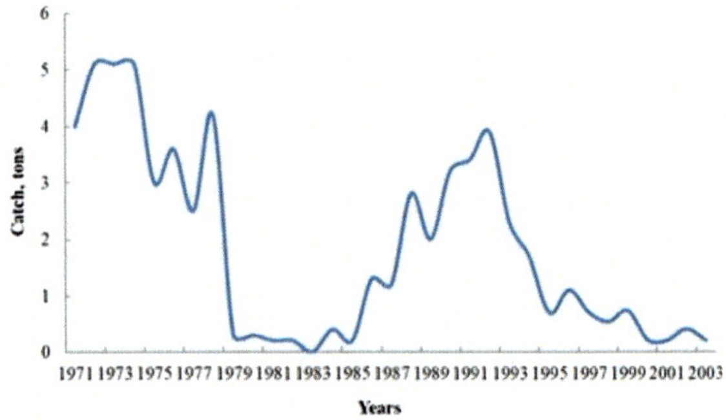

Fig. 13.2 Dynamics of commercial catches of sterlet within the boundaries of the Cheboksary reservoir. *Source* Compiled by the authors

the biomass index, a weighted average was used, taking into account the concentration of fish in both water areas.

An analysis of the dynamics of sterlet biomass indices shows their sharp decline since 1993 (see Fig. 13.3), which coincides in time with the decline in catches in fishing statistics. In addition, it was found that two peaks are observed in the dynamics of sterlet biomass. The first corresponds to the fifth year (1986) after the filling of the reservoir and corresponds to the maximum biomass of sterlet in the reservoir (biomass index 1.8 kg/ha), when the spawning grounds that existed on the Volga were still functioning. Then, after a five-year decline, a second peak was observed, but already smaller (0.9 kg/ha). At this moment, apparently, there were already less productive non-silty spawning grounds, which could no longer support a high fish biomass in the reservoir. By the beginning of the 2000s, only a small number of spawning grounds remained (see Fig. 13.1), which stably maintained a low biomass of sterlet (0.02–0.06 kg/ha).

After the start of regular releases of juveniles in the reservoir, which began in 2009, approximately in the seventh year, the dynamics of sterlet biomass in the reservoir became positive (see Fig. 13.4), which indicates the success of work on the artificial reproduction of sterlet.

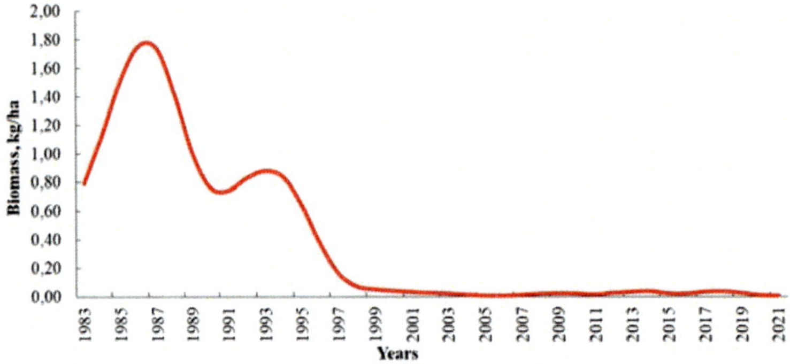

Fig. 13.3 Dynamics of sterlet biomass index in the Cheboksary reservoir. *Source* Data of the Nizhny Novgorod branch of the FSBSI "VNIRO"

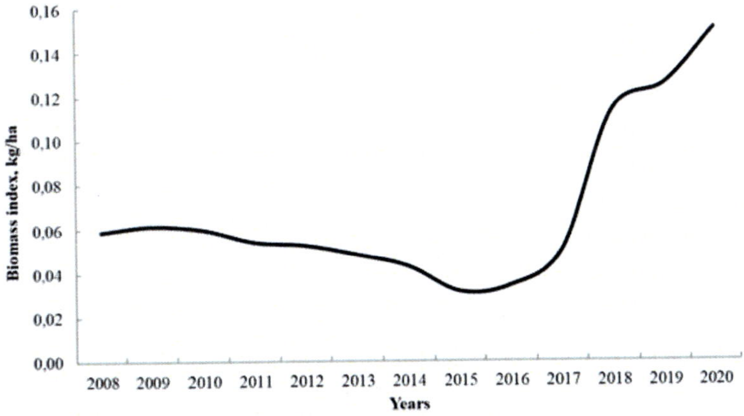

Fig. 13.4 Dynamics of the sterlet biomass index during the period of work on artificial reproduction in the Cheboksary reservoir. *Source* Data of the Nizhny Novgorod branch of the FSBSI "VNIRO"

Artificial releases of sterlet juveniles have been regularly carried out since 2009 (Table 13.1). They vary greatly in volume, but in general there is a tendency to increase the number of juveniles released. The volumes depend on funding from the environmental fund, which transfers fines for damage caused to water bodies in the Volga basin.

Table 13.1 Dynamics of sterlet juvenile release volumes into the Cheboksary reservoir

Years	Fingerling release, thousand specimens
2009	2.2
2010	11.3
2011	26.5
2012	50.9
2013	116.4
2014	37.8
2015	196.2
2016	39
2017	54
2018	336.1
2019	954.5
2020	246.3
Total	2071.2
Average	172.6

13.3.3 Brief Description of the Biological Characteristics of the Cheboksary Reservoir Sterlet

The rate of linear growth of sterlet in the Cheboksary reservoir is currently the highest among the reservoirs of the Middle and Upper Volga, starting from the age of four (Table 13.2). At younger ages, it is quite high, but not maximum. The sterlet of the Oka river, caught above the backwater zone of the Cheboksary hydroelectric complex, has a slightly lower growth rate.

Nowadays compared to the predamming period (Afanasyev 1985), the sterlet growth rate increased significantly starting from the second year. If in the late 1970s at the age of eight, the sterlet reached 39.4 cm, then according to our research in 2000. Its size is 48.2 cm.

The weight growth of the sterlet from the Cheboksary reservoir is fairly well (R2 = 0.95) described by the power Eq. (13.1)

$$W = 0.0052\, L^{3.0973} \tag{13.1}$$

where W—mass of fish, g;

L—fork length, cm.

The maturation of the sterlet begins at the age of five, and by the age of 14, immature individuals are practically not found. Under the conditions of the unregulated Volga, maturation ended by the age of 15 (Afanasyev 1985).

The sterlet of the Cheboksary reservoir is characterized by low absolute fecundity from 4 to 140 thousand units eggs. The relative working fecundity of sterlet females

Table 13.2 Linear growth of sterlet in various reservoirs and rivers of the Volga basin (sm)

Reservoirs and rivers	1	2	3	4	5	6	7	8
Oka river (2019–2021)	12.9	22.6	29.2	35.5	38.5	42.5	42.3	52.2
Cheboksary reservoir (2005–2021)	14.9	26.1	33.1	39.4	45.2	49.5	53.9	54.9
Volga river before the construction of the Cheboksary reservoir (1977–1978) (Afanasyev 1985)	19.2	22.8	27.8	32.1	34.6	36.6	39.1	39.4
Kuibyshev reservoir (2012–2019) (Granin et al. 2020)	20.4	26.4	30.7	34.4	37.3	39.8	42.1	44.0
Kuibyshev reservoir (1932–1933) (Granin et al. 2020)	15.6	23.4	28.4	31.8	36.3	40.2	43.2	46.0
Kuibyshev reservoir (1952–1955) (Granin et al. 2020)	18.7	24.5	29.7	35.5				
Kuibyshev reservoir (1956–1959) (Granin et al. 2020)	16.8	25.8	30	33.4	36.8	39.4	43.6	45.8
Kama river (Granin et al. 2020)	23.2	29.6	33.6	37.2	40.5	43.4	45.3	47.2
Ivankovsky reservoir (The Upper Volga) (Kuzmina 2005)	12.5	21.7						

is about 15,000 eggs per 1 kg of live weight. Under the conditions of the studied reservoir, spawning takes place from May 5 to May 25 at a water temperature of 10–13 °C. Re-maturation of females occurs in a year, males every year. The sexual structure of the stock changes with age. An increase in the proportion of females in older age groups is noted.

13.3.4 The Comparison of Efficiency of the Sterlet Natural Spawning and the Results of Its Artificial Reproduction in the Cheboksary Reservoir

Studies of downstream passage of sterlet larvae from the spawning grounds on the Oka river were carried out in 2001. The water area of this river is currently the main spawning grounds for sterlet. The number of places suitable for natural reproduction by this species of fish in 2001–2009 is assumed constant as evidenced by the consistently low biomass of sterlet in water bodies over the indicated period (see Fig. 13.3). In addition, we neglect insignificant areas of sterlet spawning grounds above the confluence of the Oka river.

According to the results of these studies, the average concentration of larvae in 1 m^3 of water was determined—0.0047 specimens. Water consumption in the Oka river for the passage period (the last decade of May) was 10,600 million m^3. Accordingly, the total number of drifted larvae will be 49.8 million specimens.

For fish larvae, according to the Methodology (Ministry of Agriculture of the Russian Federation 2020), the commercial return rate is 0.01%. Then there will be 4982 specimens of commercial size at the output

$$49.82 \text{ million sp.} \times 0.01/100 = 4982 \text{ sp.}$$

The statistics of sterlet juveniles releases that began regularly since 2009 and massively since 2015 were given above (see Table 13.1). The average release was 172,596 specimens of juveniles with an average weight of 10 g.

For sterlet juveniles weighing 10 g, according to the Methodology (Ministry of Agriculture of the Russian Federation 2020), the fishery return rate is 6.7%. Then individuals of commercial size, according to formula (13.1), will be 11,564 specimens in the output

$$172596 \text{ sp.} \times 6.7/100 = 11564 \text{ sp.}$$

The fishing measure for sterlet according to the Fishing Rules (Ministry of Agriculture of the Russian Federation 2019) is 42 cm. In this case, the weight of the fish will be 0.55 kg. The annual commercial return from the artificial reproduction of sterlet in terms of weight will be 6.4 tons, the return from the rolling larvae from the few remaining on the river. Oka spawning grounds—2.7 tons.

Thus, the commercial return from the artificial reproduction of sterlet is 2.3 times higher than the natural reproduction of this species of fish.

13.4 Discussion

Recent studies on the ecology and breeding characteristics of sterlet were carried out even before the creation of the Cheboksary reservoir, in the conditions of the unregulated Volga and in the first years of its existence (Afanasyev 1980, 1985, 1987; Afanasyev and Shurukhin 1987; Shibaev et al. 1989).

Based on the experience of operating previous reservoirs, and the Cheboksary reservoir was the last closing link in the Volga cascade, Afanasiev Y. I. back in 1980 (Afanasyev 1980) wrote that in the near future the sterlet may become a protected species rather than a valuable commercial species. The author pointed out that when creating artificial technogenic reservoirs on the site of a natural river system, it is necessary to create spawning and nursery farms to form commercial fish stocks.

Studies of older reservoirs, such as the Kuibyshevskoe, confirmed his assumptions. In the works of Lukin A. V., Tairov R. G., Kalaida M. L., Granin A. V. (Lukin et al. 1971; Tairov et al. 2016; Kalayda et al. 2019; Granin et al. 2020) indicates the catastrophic disappearance of the natural spawning grounds of sterlet, followed by a sharp decline in the stocks of this type of valuable biological resources. The listed researchers and many others agree that the artificial reproduction of sterlet is the only way to keep sterlet stocks at a safe level.

Despite the fact that the problem of artificial reproduction has long been known, its solution has been put off for a long time due to political and financial difficulties in the Russian Federation. Currently, they are trying to solve it by attracting funding from the environmental fund, which transfers fines for damage caused to water bodies in the Volga basin.

An analysis of the available information on the results of releases of sterlet juveniles into the Cheboksary reservoir, given in this study, shows that artificial reproduction of this fish species, which is still small in scale, already makes it possible to restore its stock.

Consequently, in order to transfer the sterlet from the category of a protected species to a commercial one, it is necessary to increase the releases of its juveniles. These activities will contribute to sustainable fishing in the Cheboksary reservoir.

13.5 Conclusion

During the 40 years of the Cheboksary reservoir existence, the main sterlet spawning grounds were irretrievably lost, which were sections of the Volga River with high flow rates and had a pebble substrate. Currently, there are several small unproductive spawning grounds in the upper part of the reservoir and the Oka River.

The loss of the main sterlet spawning grounds has led to a catastrophic decline in the stock of this valuable fish species since the mid-1990s. The biomass of sterlet decreased by 200 times and reached its minimum levels in the early 2000s. At present, a insignificant level of natural reproduction keeps the sterlet stock at a low level.

Reservoir habitat conditions led to an increase in the rate of linear and weight growth of sterlet.

The regular releases of sterlet juveniles, which began in 2009, made it possible to improve the state of the stock. The biomass of this fish species has risen five times over the past five years. However, the scale of artificial releases is small, although they have become regular in the last 10 years.

Comparison of the levels of coexisting natural and artificial reproduction of sterlet showed the prevalence of the latter, i.e. The main role in the state of the stock of sterlet in the Cheboksary reservoir is currently played by artificial releases of juveniles of this valuable species of aquatic biological resources.

In the future, it is necessary to study the migration, genetic and morphological and physiological characteristics of the sterlet of the Cheboksary reservoir and the Oka River in order to properly manage the process of restoring the stock of this valuable species of aquatic biological resources.

Thus, a further increase in artificial releases of sterlet juveniles based on the positive experience gained will contribute to the creation of a sustainable fishery in the region.

References

Afanasyev YI (1980) Breeding biology of sterlet in the unregulated Volga in the area of the future Cheboksary reservoir. Sbornik Nauchnykh Trudov GosNIORKH 157:40–49 (in Russian)

Afanasyev YI (1985) Patterns of variability in the growth of sterlet in the unregulated Volga in the area of the Cheboksary reservoir before its formation. Sbornik Nauchnykh Trudov GosNIORKH 240:73–85 (in Russian)

Afanasyev YI (1987) Duration of the sexual cycle of sterlet females under conditions of the unregulated Volga. Sbornik Nauchnykh Trudov GosNIORKH 267:47–61 (in Russian)

Afanasyev YI, Shurukhin AS (1987) Morphological characteristics of the Volga and Oka sterlet inhabiting the area of the Cheboksary reservoir in the period before its formation. Sbornik Nauchnykh Trudov GosNIORKH 267:62–80 (in Russian)

Baturina NS (2019) Regularities of the organization of river ecosystems: a retrospective of the formation of modern concepts (review). Inland Water Biol 1:3–11 (in Russian)

Boldina IK (1962) Some features of the ecology and nutrition of sterlet in the Gorky and Kuibyshev reservoirs. Iss Ichthyol 5:12–13 (in Russian)

Bondarchuk OL (2016) Formation of adaptive behavior of young sterlets (Acipenserruthenus L.) depending on the duration in breeding pools (Dissertation of Candidate of Biological Sciences). IBIW RAS, Borok, Russian Federation (in Russian)

Bykov AD (2003) Biology and artificial reproduction of sterlets of the Upper Oka (Dissertation of Candidate of Biological Sciences). FSUE "VNIIPRKH", Moscow (in Russian)

Górski K, van den Bosch LV, van de Wolfshaar KE, Middelkoop H, Filippov OV, Zolotarev DV, Verreth JAJ (2012) Post-damming flow regime development in a large lowland river (Volga,

Russian Federation): implications for floodplain inundation and fisheries. River Res Appl 28:1121–1134

Granin AV, Shakirova FM, Tairov RG, Gorshkov MA, Kalayda AE, Anokhina OK, Valieva GD (2020) Growth of Sterlet (Acipenser ruthenus L.) in Kuibyshev reservoir (according to materials 2012–2019). Series: Fishing industry, Vestnik of Astrakhan State Technical University, vol 3, pp 40–49 (in Russian)

Guti G, Berczik Á (2014) Criteria of sustainable management of large river systems—ecological aspects and challenges of the 21st century. Opusc. Zool. Budapest, 45(1), 95–99. Technical University. Series: Fishing Industry 3:40–49

Kalayda ML, Abdrakhmanov IK, Khamitova MF, Kalayda AA (2019) Release of sterlet (lat. Acipenser ruthenus) in the Kuibyshev reservoir is an important task for the development of aquaculture. IOP Conf Series: Earth Environ Sci 288:012044. https://doi.org/10.1088/1755-1315/288/1/012044 (in Russian)

Kalmykov VA (2005) Migration, distribution, population structure and stocks of sterlet (Acipenser ruthenus L.) in the Lower Volga (Dissertation of Candidate of Biological Sciences). FSUE "VNIIPRKH", Moscow (in Russian)

Khoroshko PN, Vlasenko AD, Novikova AS (1971) Atlas of sturgeon spawning grounds in the Volga basin. Offset Printing Factory, Volgograd, USSR (in Russian)

Kuzmina LA (2005) Fish-breeding biological assessment of the introduction of sterlet into the Ivankovskoe reservoir (Dissertation of Candidate of Biological Sciences). FGUP "VNIIPRH", Moscow (in Russian)

Lukin AV (1941) On the stages of sexual maturity in starlet. Doklady AN SSSR 32(5):374–376 (in Russian)

Lukin AV, Smirnov GM, Platonova OP (1971) Fish of the Middle Volga region. KSU, Kazan, USSR (in Russian)

Mamontov YP, Gepetskiy NE (2001) Fundamentals of rational sturgeon breeding in Russia. Fish Farm Fish 1:14–16 (in Russian)

Meisner VI (1933) Commercial ichthyology. Snabtekhizdat, Moscow, USSR (in Russian)

Middelkoop H, Schoor MM, Babich DB, Alabyan AM, Shoubin MA, van den Berg JH, Dijkstra J (2005) Bio-morphodynamics of the lower Volga river—a reference for river rehabilitation in The Netherlands. Archiv fur Hydrobiologie Large Rivers 15:89–103

Mikhailova MV (2001) Artificial reproduction of sturgeon in the Volga-Caspian region: yesterday, today, tomorrow. Fish Farm Fish 1:45–46 (in Russian)

Ministry of Agriculture of the Russian Federation (2019) Order of the Ministry of Agriculture of Russia «Ob utverzhdenii pravil rybolovstva dlya Volzhsko-Kaspiyskogo rybokhozyaystvennogo basseyna» (July, 25, 2019 No. 438). Ministry of Agriculture of the Russian Federation, Moscow (in Russian)

Ministry of Agriculture of the Russian Federation (2020) Order of the Ministry of Agriculture of Russia «Ob utverzhdenii metodiki ischisleniya razmera vreda, prichinennogo vodnym biologicheskim resursam» (March, 31, 2020 No. 167). Ministry of Agriculture of the Russian Federation, Moscow (in Russian)

Schletterer M, Shaporenko SI, Kuzovlev VV, Minin AE, Van Geest GJ, Middelkoop H, Górski K (2019) The Volga: management issues in the largest river basin in Europe. River Res Appl 35:510–519

Shibaev SV, Lysenko NF, Postnov DI (1989) Age structure of the population and natural mortality of sterlet in the Cheboksary reservoir. Sbornik Nauchnykh Trudov GosNIORKH 290:93–99 (in Russian)

Siberian scientists for the first time chipped sturgeon fry (2019). https://news.sfu-kras.ru/node/22064 (in Russian)

Tairov RG, Shakirova FM, Severov YA, Kalayda AE, Gorshkov AM (2016) The current state of the Kuibyshev reservoir sterlet, opportunities and challenges for restoration and maintenance of its stocks. In: Lukin AA, Pedchenko AP (eds) Proceedings from VC: the current state of biological resources of inland water bodies and ways of their rational use. FSBSI "GosNIORKH", Kazan (in Russian)

Torabi Haghighi A, Klove B, Marttila H (2014) Development of a new index to assess river regime impacts after dam construction. Glob Planet Change 122:186–196

WWF (2018) Living planet report—2018: aiming higher. In: Grooten M, Almond REA (eds). WWF, Gland

Zalepukhin VV (2009) The value of artificial reproduction in the preservation of biological diversity and replenishment of commercial fish stocks. In: Negrobov OP (ed) Modern problems of biodiversity. VSU, Voronezh, pp 142–149 (in Russian)

Chapter 14
Dynamics of Water Levels in the Gorky Reservoir (Russian Federation) During the Spawning Period and Its Influence on the Area of Spawning Grounds and Natural Reproduction of Limnophilic Fish Species

L. M. Minina and A. E. Minin

Abstract On the base of geographic information systems (GIS) the relationship between the areas of spawning areas of the Gorky reservoir and water levels were analyses, as well as assess the influence of the dynamics of spawning areas on the efficiency of natural reproduction of limnophilic fish species. The boundaries of 54 spawning areas of the Gorky reservoir have been digitized and georeferenced using Landsat satellite images. For the first time for the Gorky reservoir, the areas of spawning grounds were calculated at different water levels and formulas were obtained for the dependence of the areas of spawning grounds on the water level. It is calculated that the area of spawning grounds for the lower river, lakes, and dam sections of reservoir at different water levels can change by 1.2 times, the area of the Kostroma spill—by 1.7 times. The dynamics graph of the water levels of the Gorky reservoir in the flood in years of different water content, as well as the interannual dynamics of spawning areas and the number of fingerlings are presented. It is shown that the course of the water level in the Gorky reservoir during the spawning period is usually favorable for the processes of natural reproduction of fish due to the stability of the levels during the spawning period. No correlation was found between the degree of flooding of the spawning grounds and the number of fingerlings, which can be explained by the relative constancy of water levels during spawning in the interannual dynamics, or by the insufficient volume of materials from the fry surveys.

Keywords Gorky reservoir · Water level · Landsat satellite images · Spawning grounds · Efficiency of natural fish reproduction

L. M. Minina (✉) · A. E. Minin
Nizhny Novgorod Branch of Russian Federal Research Institute of Fisheries and Oceanography, Moskovskoe Shosse St. 31, 603116 Nizhny Novgorod, Russian Federation
e-mail: lmminina@yandex.ru

A. G. Arkhipov (ed.), *Sustainable Fisheries and Aquaculture: Challenges and Prospects for the Blue Bioeconomy*, Environmental Science and Engineering, https://doi.org/10.1007/978-3-031-08284-9_14

14.1 Introduction

Impact of regulation of the Volga river on the state of aquatic biological resources turned out to be predominantly negative (Chavychalova and Fomin 2018; Shakirova et al. 2012). The main reason that prevents the formation of significant commercial fish stocks in reservoirs is considered by many authors to be an unfavorable level regime, leading to a reduction in flooded spawning areas and, as a consequence, a decrease in the efficiency of natural fish reproduction (Chavychalova 2013; Shakirova et al. 2012; Stolbunov 2016; Vasilchenko and Ponomarev 2017; Vasiliev et al. 2016). It was determined that the flooded areas of spawning grounds can vary several times depending on the water level (Katunin 2014; Vasiliev et al. 2016). In this regard, it is relevant and valuable to obtain quantitative data on the influence of the level regime of reservoirs on the size of spawning areas and on the efficiency of natural reproduction of fish in reservoirs.

The Gorky reservoir, located on the border of the Upper and Middle Volga, was formed as a result of the regulation of the Volga river flow by the dam of a hydroelectric power station, filling was completed in 1957.The area at a normal retaining level (84 m of the Baltic system) is 144.8 thousand hectares (Minina and Minin 2021). The reservoir belongs to the type of plain-river water bodies with seasonal flow regulation (Avakyan et al. 1979). The water level depends on the surface runoff and the operating mode of the Nizhegorodskaya hydroelectric power station, as well as the upstream Rybinsk hydroelectric complex. The area of shallow waters from 0 to 2 m is 45.5 thousand hectares (31%). The Gorky reservoir is a fishery reservoir of regional significance for four constituent entities of the Russian Federation (Yaroslavl, Kostroma, Ivanovo and Nizhny Novgorod regions). Up to 1000 tons of fish are caught annually in the reservoir, of which the catch of limnophilic fish species is about 76%. Since 2008, there has been a tendency to increase the catch volume of limnophilic fish species and, accordingly, the total fish catch (Nizhny Novgorod branch of VNIRO 2021).

According to the nature of the hydrological regime and morphometry in the Gorky reservoir, 5 sections are distinguished (from top to bottom along the Volga river): the upper river, Kostroma spill, middle river, lake and dam. The upper river section is located from the Rybinskaya HPP dam to the city of Kostroma. The hydrology and morphology of the site have changed little compared to the period before the formation of the reservoir, the shallow water zone is limited. The Kostroma spill is a lake-type reservoir formed as a result of the flooding of the Kostroma lowland, it contains a significant proportion of shallow waters from 0 to 2 m (80%). The shallow zone (depths from 0 to 2 m) of the mid-river section, stretching from the city of Kostroma to the mouth of the Yolnat river inclusive, is very limited (5.5 thousand hectares, 25.4%). The lake section is located from the mouth of the Yolnat river to the settlement of Katunki, Nizhny Novgorod region, has significant extensions in river mouths. The area of its shallow waters is 17.3 thousand hectares (23.9%). The dam section is characterized by fluctuations in the water level depending on the releases of the Gorkovskaya HPP; the shallow waters in it are 2.1 thousand hectares (11.3%) (Postnov 2013).

The middle river, lake and dam sections are of great importance for the reproduction of fish. Here there is about 70% by the number (63% by area) of the spawning sites of the reservoir. Spawning grounds in this part of the water area are floodplain coastal areas, bays, estuarine areas of rivers, as well as shallow waters around the islands and off the coast. By the nature of the water regime and the development of forage aquatic organisms, the Kostroma spill is an extremely important area for the reproduction and feeding of juveniles and adult fish of the Gorky reservoir.

Geographic information systems (GIS) are used to determine the boundaries of spawning areas, their digitization and geographic referencing, as well as to determine their areas (Kalyuzhnaya e al. 2017; Komarova and Filonenko 2015; Tyurnin 2007; Zolotukhin and Hodger 2007). Mapping of spawning grounds in GIS is most often carried out using data from remote sensing of the Earth from space (space images). The advantage of satellite images is their availability and the possibility of high measurement accuracy (Kalyuzhnaya et al. 2017). However, works on comparing data on water levels from hydrological observation posts and areas of spawning grounds determined from satellite images are extremely rare and absent for the Gorky reservoir.

14.2 Materials and Methods

The aim of the work was to establish, using satellite images, the relationship between water levels and areas of spawning grounds of the Gorky reservoir in the spring–summer period, as well as to assess the impact of changes in spawning areas on the amount of replenishment of limnophilic fish species, forming the basis of ichthyocenosis. To achieve the goal, the following tasks were set:

- search and selection of Landsat satellite images;
- GIS mapping of spawning areas by the method of interpretation of satellite images;
- joint analysis of diverse data (numerical—values of water levels and spatial—digitized spawning areas);
- identification of factors influencing the efficiency of natural reproduction of limnophilic fish species of the Gorky reservoir.

With the help of GIS, the boundaries of the spawning grounds of the Gorky reservoir were digitized and geographically linked (see Fig. 14.1), while the list of spawning grounds in Appendix No. 6 to the Fishing Rules for the Volga-Caspian fishery basin (2019) was taken as a basis. Additionally, the analysis included spawning grounds recommended for inclusion in the list of spawning grounds of the Fishing Rules by departments of state control, supervision and protection of aquatic biological resources in Ivanovo and Nizhny Novgorod regions.

Mapping of spawning grounds was carried out by the method of visual interpretation and digitization of the coastline using satellite images from Landsat 5, Landsat

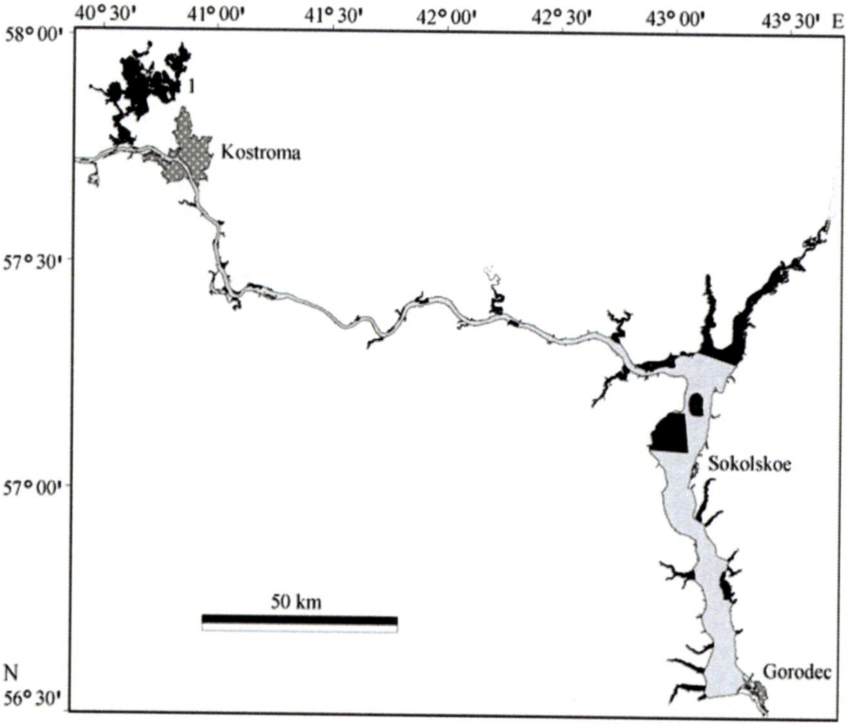

Fig. 14.1 Location of spawning areas of the Gorky reservoir (Kostroma spill, middle river, lake and dam sections): gray solid fill—Gorky reservoir, black fill—spawning grounds (1—Kostroma spill). *Source* Compiled by the authors

7 and Landsat 8 satellites (the resolution of channels in the visible part of the spectrum is 30 m/pixel). The boundaries of the spawning grounds included areas with depths from 0 to 3 m; the depths were determined using a vector map of the depth distribution of the Gorky reservoir (Nizhny Novgorod branch of VNIRO 2013).

To digitize the boundaries of spawning areas, satellite images from the USGS Global Visualization Viewer (2021) were used. The material for the study was 23 sets of Landsat satellite images for the water area of the lower river, lake and dam sections of the Gorky reservoir for 2004–2009, 2011, 2014–2020 during the spawning period (from April 8 to June 15). In the absence of a space image that completely covers the studied area, pairs of images were used for the same date, or differing in time of creation by 1–6 days, while the difference in water levels on the date of creation of the images was no more than 6 cm. 2 space images were analyzed and 3 sets of two images of the lower river, lake and near-dam areas, taken in April, 4 images and 9 sets of two images—in May, 3 images and 2 sets of two images—in June.

The Kostroma spill, which has peculiar conditions (features of origin and hydrological regime, predominance of shallow waters, exceptional feeding possibilities)

(Zelikman 1965), was considered separately from other areas. To assess the areas of the Kostroma spill at different water levels, 28 satellite images were used for the spawning period of 2002, 2004–2011, 2013–2019. We used 8 satellite images taken in April, 13 images—in May and 7 images—in June.

The water levels for the lower river, lake and dam sections are taken according to the data of the hydrological station in the settlement of Sokolskoye in the Nizhny Novgorod region, for the Kostroma spill—according to the data of the station in the city of Kostroma. The data on water levels corresponding to the time of creation of each satellite image was obtained from the archive located in the portal of the Center for Registry and Cadastre (2021). When the water levels differed on the dates of the photographs of one pair, their values were averaged.

Materials on quantitative indicators of reproductive success of limnophilic fish species (bream *Abramis brama* (L., 1758), pike *Esox lucius* L., 1758, roach *Rutilus rutilus* (L., 1758), common carp *Cyprinus carpio* L., 1758, asp *Leuciscus aspius* (L., 1758), crucian carp *Carassius auratus* (L., 1758), rudd *Scardinius erythrophthalmus* (L., 1758), tench *Tinca tinca* (L., 1758), silver bream *Blicca bjoerkna* (L., 1758), blue bream *Ballerus ballerus* (L., 1758)) were collected during seine surveys (by drag and fry seine) in the studied sections of the Gorky reservoir in 2002, 2006, 2010, 2011, 2014, 2016–2020.

14.3 Results

The graph of changes in the water levels of the Gorky reservoir during the spawning period in years with different water availability is shown in Fig. 14.2.

Based on the results of decoding the Landsat satellite images, they digitized and linked to the geographic grid of coordinates of the border of 53 spawning areas of the middle river, lake and dam sections of the Gorky reservoir, as well as the Kostroma spill. 23 variants of the area of spawning grounds of the middle river, lacustrine and near-dam sections were determined at different water levels, 28 variants of the area of the Kostroma spill (see Fig. 14.3).

The regression dependence of the area of spawning grounds in the middle river, lake and dam sections of the Gorky reservoir on the water level was obtained:

$$y = 2360.5858 - 58.6976\,x + 0.3693\,x^2 \tag{14.1}$$

where: y is the area of spawning grounds, thousand hectares;

x is the water level at the Sokolskoye hydrological station, m.

The dependence of the area of the Kostroma spill on the water level is expressed by the following formula:

$$y = 24016 - 576.37\,x + 3.4603\,x^2 \tag{14.2}$$

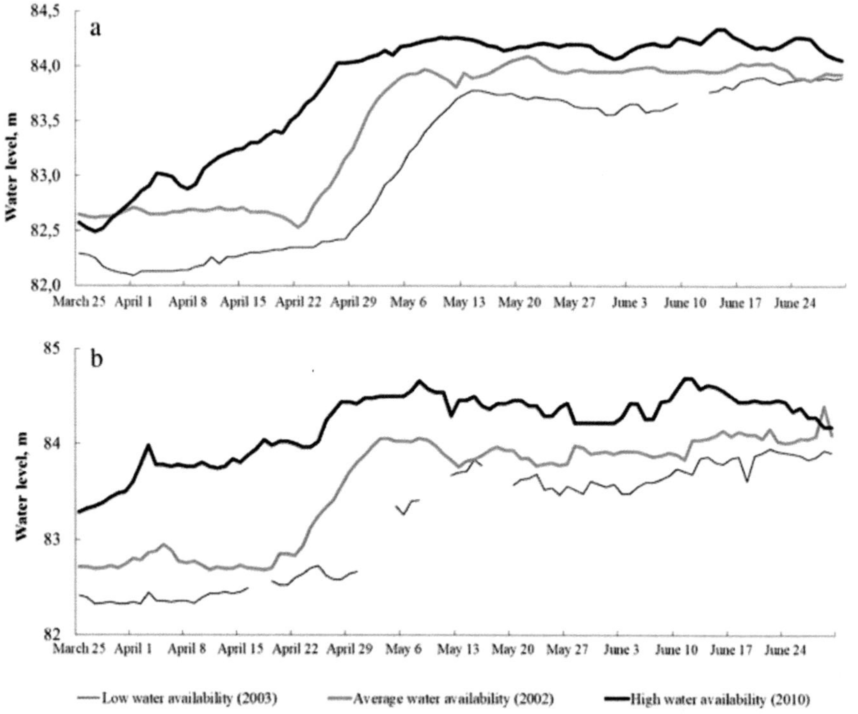

Fig. 14.2 The dynamics of the water levels of the Gorky reservoir in the spring–summer period in the years of different water availability: **a** according to the data of the hydrological station in the settlement of the town of Sokolskoye, **b** according to the data of the hydrological station in the city of Kostroma. *Source* Compiled by the authors

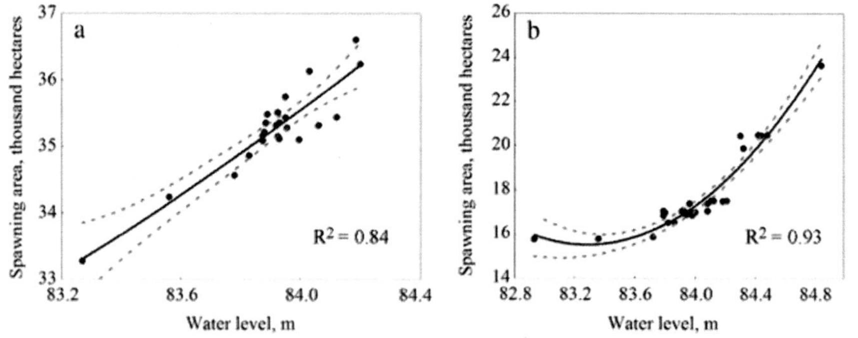

Fig. 14.3 Dependence of the area of spawning areas of the Gorky reservoir on the water level: **a** middle river, lake and dam sections (n = 23), **b** Kostroma spill (n = 28): solid line—regression curve, dotted line—borders of 95% confidence interval for regression curve. *Source* Compiled by the authors

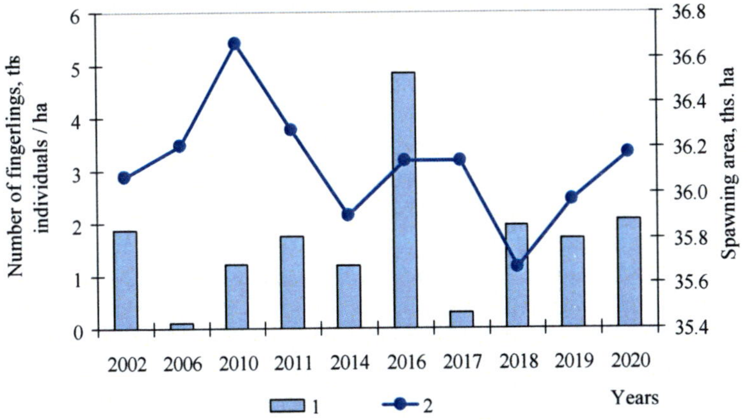

Fig. 14.4 Interannual dynamics of the relative number of fingerlings (1) and the maximum area of spawning grounds (2) of the middle river, lake and dam sections of the Gorky reservoir. *Source* Compiled by the authors

where: y is the area of the Kostroma spill, thousand hectares; x is the water level at the Kostroma hydrological station, m.

The interannual dynamics of the areas of spawning grounds and the number of fingerlings in the shallow waters of the Gorky reservoir is shown in Fig. 14.4.

Evaluation of the impact of the flooding of the Kostroma spill on the yield of fish generations is still not possible due to the small number of measurements of quantitative indicators of the success of fish reproduction.

14.4 Discussion

According to the graph in Fig. 14.2, the dynamics of the water levels of the Gorky reservoir in the flood in different water years is similar. High water in the Gorky reservoir usually begins in early April (median—April 4). After reaching a peak in early May (median—May 4), no noticeable decrease in water level until the winter low-water period, which was also noted by Kozhevnikov (1965, 1974). The absence of a sharp drop in the water level after the peak of high water (see Fig. 14.2) ensures a sufficient duration of flooding of spawning grounds and stable conditions for the development of eggs and juveniles of fish, which is reflected in the stability of the relative abundance of fingerlings in the interannual dynamics (see Fig. 14.4). For other sections of the Volga river it was also noted that the decline rate of the flood wave with an insufficient duration of flooding of spawning grounds is of decisive importance for the productivity of spawning grounds (Taradina and Chavychalova 2017; Zhezmer and Buber 2018).

Analysis of the degree of flooding of spawning grounds, carried out using satellite images and a regression model (14.1), showed that the total area of spawning grounds in the middle river, lake and dam sections of the Gorky reservoir in the spring varies depending on the water level within the following limits: minimum—30.92 thousand hectares, maximum—37.02 thousand hectares (at water levels, respectively, 82.19 and 84.36 m; the minimal and maximal water levels are determined according to the Center of the Register and Cadastre for 2002–2021). The possible range of variation in the spawning areas of the middle river, lake and dam sections of the Gorky reservoir in the spring is 1.2 times. The areas of the Kostroma spill can vary 1.7 times: the minimum is 15.8 thousand hectares, the maximum is 26.3 thousand hectares (at water levels of 82.93 and 85.09 m, respectively). Kozhevnikov (1965) indicated the area of the Kostroma spill equal to 26 thousand hectares.

No correlation was found between the number of fingerlings of limnophilic fish species in the middle river, lacustrine, and dam sections of the reservoir and the maximum area of spawning grounds. In contrast to the Gorky one, for the Cheboksary reservoir, a noticeable positive relationship was established between the number of fingerlings and the average area of spawning grounds—the Spearman correlation coefficient is 0.65 at $p = 0.01$ (Minina et al. 2021). Perhaps this is due to the insignificant range of variation in the areas of spawning grounds of the Gorky reservoir at different water levels (by a factor of 1.2). The spawning grounds of the Cheboksary reservoir can increase in area by 1.7 times (Minina et al. 2021). This difference is probably determined by the fact that the area of the Gorky reservoir is characterized by less variability depending on the water level, in contrast to the Cheboksary one (Minina and Minin 2021). Also, one of the reasons for the lack of correlation between the number of juveniles and the area of spawning grounds in the Gorky reservoir may be the insufficient amount of data on juvenile surveys ($n = 10$). In the work of Klevakin with co-authors (2012), it is indicated that at different values of the maximum water level of the Gorky reservoir, the total number of fingerlings can differ by 36–211 times.

Further research can help to obtain a formula that allows predicting the amount of natural reproduction of fish based on known water levels and areas of spawning areas determined from them, which will allow forecasting of fish stocks and will help to determine the optimal values of the total allowable catch and recommended catch.

14.5 Conclusion

The total area of spawning areas of the lower river, lake and dam sections of the Gorky reservoir can increase by 1.2 times with an increase in the water level, the area of the Kostroma spill—by 1.7 times.

The obtained formulas for the dependence of the areas of spawning grounds on the water level can be used to predict the degree of flooding of spawning grounds in specific years.

Stabilization of the water level of the Gorky reservoir after reaching the peak of the flood waters ensures the preservation of the water availability of the spawning grounds, which favorably affects the success of the natural reproduction of fish. The lack of correlation between the breeding efficiency of limnophilic fish species of the river, lake and dam sections of the Gorky reservoir and the area of spawning grounds may be explained by insignificant differences in the values of spawning areas at different water levels, or by the lack of data on juvenile surveys.

The method of reconstructing spawning areas from satellite images in coordination with data from hydrological observation posts makes it possible to compare the amount of fish recruitment directly with the areas of spawning areas, and not with water levels.

Due to the geographic referencing in GIS of the boundaries of spawning areas, discrepancies in the interpretation of their descriptions in the Fishing Rules and localization on the terrain have been eliminated, proposals have been prepared for amending the Fishing Rules, indicating the coordinates of the boundaries of the areas, which will enhance the effectiveness of their protection.

The conducted studies help to use fish stocks without harming fisheries and the environment in four regions of the Russian Federation and contribute to the development of sustainable fishing in the Gorky reservoir.

References

Avakyan AB, Saltykin VP, Sharapov VA (1979) Reservoirs. Thought, Moscow (in Russian)

Center for register and cadastre (2021). https://gis.vodinfo.ru (in Russian)

Chavychalova NI (2013) Modern problems of natural reproduction of fish in the lower reaches of the Volga. Inzhenerno-stroitel'nyj vestnik Prikaspiya (Eng Constr Bull Casp Reg) 2:80–87 (in Russian)

Chavychalova NI, Fomin SS (2018) The effectiveness of natural reproduction of fish depending on the hydrological regime of the Volga river in the spring-summer period. Rybnoe Hozyajstvo (Fisheries) 6:12–15 (in Russian)

Kalyuzhnaya NS, Kalyuzhnaya IYu, Khoruzhaya VV, Samoteeva VV, Sokhina EN (2017) Experience in studying the state of spawning grounds in the upper reach of the Tsimlyansk reservoir using GIS. Materials of the international conference: InterCarto. InterGIS 23 (InterCarto. InterGIS 23). Publishing House of Moscow State University, Moscow (in Russian)

Katunin DN (2014) Hydroecological foundations of the formation of ecosystem processes in the Caspian Sea and the Volga River delta. KaspNIRKH, Astrakhan (in Russian)

Klevakin AA, Loginov VV, Minin AE, Postnov DI (2012) Determination of irreversible river flow and establishment of ecological release in the Gorky reservoir. Water: Chem Ecol 8:8–15 (in Russian)

Komarova AS, Filonenko IV (2015) Estimation of the area of spawning grounds for phytophilous fish of lake Beloye using GIS technologies. In: Materials of XIII All-Russian scientific conference: Vuzovskaya nauka—regionu. VoSU, Vologda (in Russian)

Kozhevnikov GP (1965) Gorky Reservoir as a habitat for fish. Izvestiya Gosudarstvennogo nauchno-issledovatel'skogo instituta ozernogo i rechnogo rybnogo hozyajstva. 59:7–18 (in Russian)

Kozhevnikov GP (1974) General characteristics of the shallow zone of the Gorky reservoir. Izvestiya Gosudarstvennogo nauchno-issledovatel'skogo instituta ozernogo i rechnogo rybnogo hozyajstva. 89:14–20 (in Russian)

Minina LM, Minin AE (2021) Clarification of the areas of the Gorky and Cheboksary reservoirs on the basis of remote sensing data of the Earth. KGTU NEWS 60:44–54 (in Russian)

Minina LM, Minin AE, Moiseev AV (2021) Influence of the dynamics of water level in spring on the area of spawning grounds and efficiency of natural reproduction limnophilic fish species of the Cheboksary reservoir. Trudy VNIRO 185:84–93 (in Russian)

Ministry of Agriculture of Russia (2019) Order of the Ministry of Agriculture of Russia «On approval of fishing rules for the Volga-Caspian fishery basin» (July, 25, 2019 No. 438). Moscow (in Russian)

Nizhny Novgorod branch of VNIRO (2013) Research report: spawning grounds for commercial fish in the Gorky reservoir. Nizhny Novgorod branch of VNIRO, Nizhny Novgorod (in Russian)

Nizhny Novgorod branch of VNIRO (2021) Research report: materials substantiating the total allowable catch of aquatic biological resources in the Gorky reservoir (within the boundaries of the Nizhny Novgorod, Ivanovo, Kostroma and Yaroslavl regions) and water bodies of the Kostroma region for 2022. Nizhny Novgorod branch of VNIRO, Nizhny Novgorod (in Russian)

Postnov DI (2013) Regularities of the formation and rational use of biological resources of the Gorky reservoir (Dissertation of Candidate of Biological Sciences). Kaliningrad State Technical University, Kaliningrad (in Russian)

Shakirova FM, Tairov RG, Severov YuA (2012) Influence of the level regime of the Kuibyshev reservoir on the formation of its fish stocks. Fisheries 1:40–43 (in Russian)

Stolbunov IA (2016) Distribution, species composition and number of juvenile fish in the shallow-water zone of the Rybinsk reservoir in years different by the level regime. Bull Astrakhan State Tech Univ Ser: Fish 2:64–70 (in Russian)

Taradina DG, Chavychalova NI (2017) On the natural reproduction of semi-anadromous and some river fish species in the lower reaches of the Volga river in 2011–2015. Trudy VNIRO 166:85–108 (in Russian)

Tyurnin VB (2007) Application of satellite space information for the study of spawning grounds of the Okhotsk herring. In: Chereshnev IA (ed) All-Russian scientific conference: Chteniya pamyati akademika K. V. Simakova. Magadan (in Russian)

USGS Global Visualization Viewer (2021). http://glovis.usgs.gov.

Vasilchenko OM, Ponomarev VS (2017) On the features of natural reproduction of bream (Abramis brama orientalis L.) in the spawning grounds of the lower reaches of the Volga river in high-water 2013 and low-water 2015. Bull Astrakhan State Tech Univ Ser: Fish 3:17–24 (in Russian)

Vasiliev YuS, Maslikov VI, Shilin MB (2016) The flow regulation regime of the Rybinsk reservoir as the main factor in the formation of the ecological situation in the dry zone. Uchyonye Zapiski RGGMU 45:28–42 (in Russian)

Zelikman AL (1965) Kostroma spill of the Gorky reservoir as a fishery reservoir. Izvestiya Gosudarstvennogo nauchno-issledovatel'skogo instituta ozernogo i rechnogo rybnogo hozyajstva. 59:207–210 (in Russian)

Zhezmer VB, Buber AL (2018) Correspondence of the modern indicators of the hydroecological regime of the lower reaches of the Volga to the basic requirements of the spawning cycle of fish in the Volga-Caspian fishing subdistrict. Prirodoobustrojstvo 5:21–30 (in Russian)

Zolotukhin SF, Hodger LCh (2007) Calculations of the area of the salmon spawning fund of the main rivers in the south of the Amur Estuary. Izvestiya TINRO 148:130–142 (in Russian)

Chapter 15
Fisheries and Aquaculture: Implementing Sustainable Development Goals

Natalia N. Yarkina and Natalia N. Logunova

Abstract The blue bioeconomy is considered as a factor of sustainable development of society through sustainable development of fisheries and aquaculture. The purpose of the study is scientific and practical re-search on the problem of sustainable development in the context of blue bioeconomy, the development of its theoretical and applied provisions and scientific tools. Based on the application of methods of analysis, synthesis and decomposition, the structure and patterns of the functioning and development of the world fisheries have been investigated. Furthermore, the Sustainable Development Goals (SDGs) in the context of a blue bioeconomy have been specified. The use of methods of statistical and comparative analysis made it possible to identify the conditions and trends in the functioning and development of the global fisheries and aquaculture sector, to compare data on the dynamics of the global catch and population of the Earth, to assess the volume and structure of production, as well as the level of employment in the global fishing and fish farming sector. Based on the world statistics and analytics, the role of the fisheries and aquaculture sector in the implementation of sustainable development goals is shown. The SDG 14 (Conserve and sustainably use the oceans, seas and marine resources for sustainable development) and the Blue Growth initiative are emphasized. The basic characteristics of the Blue Growth initiative that are relevant in the context of the blue bioeconomy are highlighted. It is proved that for the development of the fisheries and aquaculture sector, it is necessary to competently regulate fisheries and apply methods in aquaculture that contribute to the sustainable use of resources and the conservation of aquatic biodiversity, that is, to act on the basis of the sound principles of the blue bio-economy.

Keywords Sustainable development · Goals · Blue Growth initiative · Blue bioeconomy · Fisheries · Aquaculture

N. N. Yarkina · N. N. Logunova (✉)
Kerch State Maritime Technological University, St. Ordzhonikidze, 82, 298309 Kerch Republic of Crimea, Russian Federation
e-mail: natalya_logunova@mail.ru

A. G. Arkhipov (ed.), *Sustainable Fisheries and Aquaculture: Challenges and Prospects for the Blue Bioeconomy*, Environmental Science and Engineering, https://doi.org/10.1007/978-3-031-08284-9_15

15.1 Introduction

The world we live in now (fields and forests, seas and oceans, atmosphere and biosphere), suddenly from a reliable, constant basis of our life turned into a fragile creation. It is our turn now to protect it, thus ensuring the ability to eat, maintain physical health and spiritual development, that is, just to live.

Humanity has entered the time of active exploitation of natural resources, associated with the undermining of the reproductive capacity of some of them and the degradation or complete disappearance of others. We are the only ones who can stop this destructive process, creating guarantees for current life and providing future generations the right to life.

Bioeconomics is one of the means of ensuring the sustainable development of modern society. It is also recognized as a means of ensuring the viability and prosperity of subsequent generations (Bruckner et al. 2019; Lopes et al. 2021; Hasselstrom and Grondahl 2021).

One of the prerequisites for the development of the bioeconomy concept ("green" and "blue") is the need to solve and (or) mitigate the problems of imbalance in the economic, social and environmental goals of society. Neither economic growth nor social progress is possible without maintaining environmental sustainability. Thus, the inevitability of joint action to ensure economic growth, social progress and environmental sustainability also prompted the development of the bioeconomy concept.

15.2 Materials and Methods

The objective of the research is to continue scientific and practical research on the problem of sustainable development in the context of blue bioeconomy (Austen et al. 2019) and the development of its theoretical and applied status and scientific tools.

Objectives of the study: consideration of the fisheries and aquaculture sector as a subject of the implementation of sustainable development goals; analysis of the dynamics and structure of the global fisheries and aquaculture sector; highlighting the fundamental characteristics of the Blue Growth initiative, relevant in the context of the blue bioeconomy. To carry out the study, a wide methodological toolkit was used. Thus, the application of methods of analysis, synthesis and decomposition allowed solving problems related to the study of the structure and patterns of functioning and development of the world fisheries, as well as specifying the goals of sustainable development in the context of blue bioeconomy. The use of methods of statistical and comparative analysis made it possible to identify the conditions and trends in the functioning and development of the global fisheries and aquaculture sector, to compare data on the dynamics of the global catch and population of the Earth, to assess the volume and structure of production by the global fishing and fish farming sector, as well as the level of employment in this sector. On the basis of deduction and synthesis methods, conclusions were made about the prospects for sustainable development of the fisheries and aquaculture sector. Moreover, the fundamental characteristics of the Blue Growth initiative in the context of blue bioeconomy were highlighted.

15.3 Results

Please note that the first paragraph of a section or subsection is not indented. The first Sustainable development is a global goal. Thereby, ensuring this development is the challenge that requires joint, consistent efforts of countries, international organizations and their institutions aimed at creating conditions and developing specific measures for the evolution of society and maintaining harmony between the world of the people and world of animals and plants. Only coherent, rational and effective actions of the world community will make it possible to end poverty (SDG 1), end hunger, achieve food security (SDG 2), ensure health and well-being for all (SDG 3), ensure quality education (SDG 4), ensure gender equality (SDG 5), ensure availability and sustainable management of water and sanitation (SDG 6), ensure universal access to affordable energy services (SDG 7), ensure decent work and economic growth (SDG 8), develop quality, reliable infrastructure, promote sustainable industrialization and foster innovation (SDG 9), reduce inequality (SDG 10), ensure sustainable cities and human settlements (SDG 11), ensure sustainable consumption and production (SDG 12), combat climate change (SDG 13), conserve marine ecosystems (SDG 14), conserve terrestrial ecosystems (SDG 15), promote peace, justice and efficient institutions (SDG 16), build partnerships for sustainable development (SDG 17). The presented system of seventeen large-scale sustainable development goals (SDGs), originally formulated in the Millennium Declaration, formed the basis of the 2030 Agenda for Sustainable Development—a program document adopted by the UN in 2015 (FAO 2015). To implement these goals, 169 target tasks have been established. These include the tasks of transformation in technical, institutional and political spheres. The complex solution of these tasks is considered as a guarantee of achieving sustainable development of modern society.

Fishing activities within the sectoral framework of fisheries and aquaculture are integrated into the overall process of achieving the specified goals. In the preface to the report "The State of World Fisheries and Aquaculture 2020" devoted to the theme of "Measures to Improve Resilience", the FAO Director-General Qu Dongyu emphasized that "The fisheries and aquaculture sector has much to contribute to securing all the SDGs" (The State of World Fisheries and Aquaculture 2020).

The direct participation of fisheries and aquaculture economic entities in ensuring people's livelihoods and analyzing their potential in the sustainable development system made it possible to systematize and rank the SDG groups according to the degree of influence on their implementation in the global fishery sector.

The group of the most significant, real and potential contribution of fisheries to their implementation includes such sustainable development goals as:

- SDG 2: End hunger, ensure food security and improve nutrition, and promote sustainable agricultural development;
- SDG 3: Ensure healthy lifestyles and promote well-being for all at all ages;
- SDG 8: Promote sustained, inclusive and sustainable economic growth, full and productive employment and decent work for all;
- SGG 12: Ensure the transition to sustainable consumption and production patterns;
- SDG 14: Conservation and sustainable use of the oceans, seas and marine resources is a key to sustainable development.

It is recognized that "fish and fish products are not only the healthiest products on the planet, but also have the least significant impact on the environment. Therefore, they need to become much more prominent in national, regional and global food security and nutrition strategies and should play an important role in transforming food systems. Thus, it will help to erase hunger and malnutrition" (FAO).

Hunger and malnutrition is a very complex issue that requires urgent solutions. Today, one in nine people on the planet is undernourished. According to official UN statistics, 821 million people faced malnutrition in 2018. It should be noted that the number of hungry people has been decreased over the years, but this figure has been increasing since 2014.

The fisheries and aquaculture sector makes a unique contribution to improving all four dimensions of food security: availability, access, use and stability.

According to Table 15.1, the growth in dynamics of the world catch of fish and seafood in terms of live weight and the number of inhabitants of our planet during 1900–2018 is evident. It is seen in favour of fishing activities results. For more than a century, there has been the sharp growth in the production of world fisheries products in live weight equivalent compared to the growth in the world's population.

Of the total world catch in 2018 of 178.5 million tons of fish in live weight equivalent (the term "fish" includes fish, crustaceans and molluscs), 156 million tons were used for human consumption, which is equivalent to 20.5 kg per capita per year. Comparatively, in 1961 the consumption of food fish in live weight equivalent was 9.0 kg. Since 1961, per capita consumption of food fish has grown by about 1.5% per year. The rate of growth in the production of fish products in live weight equivalent significantly exceeded the rate of increase in the volume of production of all other products containing animal proteins (meat, dairy products, milk, etc.). In 2017, fish represented about 17% of animal protein and 7% of all protein consumed by people in the world (FAO).

Despite the stable differences in fish consumption in different regions and states, the following positive trends are seen (FAO):

- in developed countries, fish consumption per capita increased from 17.4 kg in 1961 to the maximum level of 26.4 kg in 2007, and then began to decrease and in 2017 amounted to 24.4 kg;
- in developing countries, fish consumption per capita in 1961 was 5.2 kg, and in 2017 amounted to 19.4 kg, which corresponds to an average annual growth rate of 2.4%;

Table 15.1 Dynamics of world catch and population of the Earth for 1900–2018

Index	Years											
	1990	1950	1960	1970	1980	1990	2000	2021	2015	2016	2017	2018
World catch, million tons	4.0	21.1	40.0	70.8	71.9	97.7	126.0	148.1	168.7	166.1	172.7	178.5
Population of the Earth, billion people	1.6	2.5	3.0	3.6	4.4	5.3	6.1	6.9	7.3	7.4	7.5	7.6
Average annual growth rate for the previous chain period, %												
World catch		3.4	6.6	5.9	0.2	3.1	2.6	1.6	2.6	−1.5	4.0	3.4
Population size		0.9	1.8	2.2	2.0	1.9	1.4	1.2	1.1	1.4	1.3	1.3

Source FAO

- in the least developed countries, the annual per capita fish consumption increased from 6.1 kg in 1961 to 12.6 kg in 2017, i.e., grew on average by 1.3% per year. In the past 20 years, the growth rate of this indicator has increased to 2.9% per year, which is associated with an increase in the production and import of fish;
- in low-income and food-deficit countries, fish consumption has grown steadily by about 1.5% per year; in 1961 it was 4.0 kg, and in 2017 it was already 9.3 kg.

The contribution of the world fishery sector to solving food problems is unquestionable, and its potential is demonstrated by the dynamics of production volumes in live weight equivalent for 2000–2018 (see Table 15.2).

In 2018, the world produced a record number of industrial fishery products—96.4 mil-lion tons. However, the possibilities for a sustainable increase in the production of fish products are not just limited, but also constrained by the factor of the natural reproductive capacity of the raw material base of oceanic fisheries, significantly

Table 15.2 The volume and structure of production by the global fishing and fish farming sector in 2000–2018

Year	World catch, in total		Including							
			Fishing, in total		Sea waters		Inland waters		Aquaculture, in total	
	Million tons	%	Million tons	%	Million tons	%	Million tons	%	Million tons	%
2000	126.0	100.0	93.6	74.3	85.0	67.5	8.6	6.8	32.4	25.7
2001	129.8	100.0	93,4	72.0	84.8	65.3	8.6	6.7	36.4	28.0
2002	133.6	100.0	93.2	69.8	84.5	63.2	8.7	6.6	40.4	30.2
2003	133.2	100.0	90.5	67.9	81.5	61.2	9.0	6.7	42.7	32.1
2004	134.3	100.0	92.4	68.8	83.8	62.4	8.6	6.4	41.9	31.2
2005	136.4	100.0	92.1	67.5	82.7	60.6	9.4	6.9	44.3	32.5
2006	137.3	100.0	90.0	65.5	80.2	58.4	9.8	7.1	47.3	34.5
2007	140.7	100.0	90.8	64.5	80.7	57.3	10.1	7.2	49.9	35.5
2008	143.1	100.0	90.2	63.0	79.9	55.7	10.3	7.3	52.9	37.0
2009	145.9	100.0	90.2	61.8	79.7	54.6	10.5	7.2	55.7	38.2
2010	148.1	100.0	89.1	60.2	77.8	52.5	11.3	7.7	59.0	39.8
2011	154.0	100.0	92.2	59.9	81.5	52.9	10.7	7.0	61.8	40.1
2012	156.0	100.0	89.5	57.4	78.4	50.3	11.1	7.1	66.4	42.6
2013	160.7	100.0	90.6	56,4	79.4	49.4	11.2	7.0	70.1	43.6
2014	164.9	100.0	91.2	55.3	79.9	48.4	11.3	6.9	73.7	44.7
2015	168.7	100.0	92.6	54.9	81.2	48.1	11.4	6.8	76.1	45.1
2016	166.1	100.0	89.6	53.9	78.2	47.1	11.4	6.8	76.5	46.1
2017	172.7	100.0	93.1	53.9	81.2	47.0	11.9	6.9	79.6	46.1
2018	178.5	100.0	96.4	54.0	84.4	47.3	12.0	6.7	82.1	46.0

Source FAO

undermined by ecosystem problems caused by the intensification of fishing because of climate change and negative anthropogenic impact.

During the period under study, the volume of catches of world fisheries fluctuated with-in the range of 89.1–96.4 million tons. Thus, there was no strong tendency of development either to increase or to decrease, which allows us to conclude that the global oceanic catch has reached its critical point. In turn, the world aquaculture during 2000–2018 developed according to an arithmetic progression, and the average annual absolute increase in the volume of production of its products in live weight equivalent was 2.79 million tons. According to Table 15.2, the volume of withdrawals of world fish farming products in 2000 was 32, 4 million tons or 25.7% of the total world catch, and in 2018—82.1 million tons or 46.0%, respectively. It was despite the fact that 52% of the fish consumed by people was produced in 2018.

The results of the study clearly indicate that the future in the development of the global fishery sector belongs to aquaculture. Aquaculture has significant potential in solving not only food problems of society (and recognized as the most important factor in global food security), but economic ones, contributing to sustained and sustainable economic growth. In particular, the ratio in 2018 of the physical volumes of the world catch of fisheries and aquaculture (in tons) was 1.17: 1 (54.0% and 46.0%, respectively), and the ratio of the market value of fishery and aquaculture products (in prices of initial sale) amounted to 0.6: 1 (37.7% and 62.3%, respectively).In absolute terms, according to 2018 data, the following is observed: total sectoral catch of fish amounted to 178.5 million tons, of which 82.15 million tons are aquaculture products; US$401 billion is the total initial sales of world fisheries products, of which US$250 billion is from aquaculture (FAO).

Sustainable development of aquaculture requires forging new strategies that will help to realize the full potential of the sector as one of the actors of global development and solve the large-scale environmental problems that will inevitably arise in the process production intensification. These strategies must take into account the technological advances in feed production, selective breeding, biosafety and disease management, as well as innovative digital solutions and new approaches to entrepreneurship in investment and trade.

The global fisheries and aquaculture sector in 2018 employed 0.8% of the world's population, i.e. almost every 125. During the period under study, there was mainly a positive trend in the number of people employed in the industry (see Table 15.3). Thus, aquaculture showed a sharp growth.

The majority of those employed in fisheries and aquaculture are from developing countries. Persons employed in artisanal, small-scale fisheries and aquaculture workers predominate. Thus, the sector addresses social and economic issues where poverty, hunger and malnutrition are most acute. Hence, the sector provides productive employment and is a source of income and food.

In the context of food deficit and limited possibilities of the raw material base of the fish industry, the problem of rational production and consumption of fish products is acute. Food loss and spoilage is a worldwide challenge.

Fish raw materials are unique as they can be fully used by processing for food and technical purposes. Modern technologies make it possible to increase the yield

Table 15.3 Level, structure and dynamics of employment in the global fisheries and aquaculture sector in 2020–2018

Fisheries and aquaculture sector	Employed, thousand people					Share, %				
	2000 year	2005 year	2010 year	2015 year	2018 year	2000 year	2005 year	2010 year	2015 year	2018 year
Fishery	34,439	36,655	39,305	38,771	38,976	72.9	70.5	67,8	65.5	65.5
Aquaculture	12,825	15,364	18,625	20,390	20,533	27.1	29.5	32.2	34.5	34.5
In total	47,264	52,019	57,930	59,161	59,509	100.0	100.0	100.0	100.0	100.0
	Average annual growth rate of the employed, %					Deviation from the previous period (±),%				
Fishery	–	101.3	101.4	99.7	100.2	–	−2.4	−2.7	−2.3	0.0
Aquaculture	–	103.7	103.9	101.8	100.2	–	+2.4	+2.7	+2.3	0.0
In total	–	101.9	102.2	100.4	100.2	–	0.0	0.0	0.0	0.0

Source FAO

of finished food products through the use of fish processing by-products, which are widely used. As we know, heads, backs, fillet trimmings and skin can be used directly in food or for preparing fish sausages, pies, pate, snacks, gelatin, soups, sauces and other foods.

Moreover, by-products are used for the production of feed and fertilizers, biofuels and biogas, dietary products (chitosan), pharmaceuticals (oils—sources of omega-3), natural dyes, cosmetics and plastic alternatives, and are also used for other industrial purposes.

By-products can make up 70% of processed fish. Consequently, it leads to serious environmental technical problems, since they have a high microbial and enzymatic load and rapidly decompose if not processed or stored them properly. Thereby, the studied issue is relevant and significant.

Sustainable use of natural resources is not only a key principle for the implementation of SDG 12, but is also a cross-cutting issue of all sustainable development goals.

The existing global sustainable development goals can be achieved only on the condition of conservation and sustainable use of the oceans, seas and marine resources (SDG 14).

15.4 Discussion

Sustainable fisheries and aquaculture is the key to achieving goals of sustainable development of modern society. As a party to the global fishery processes in the triad of the World Ocean, Subjects of fisheries and Humanity, fisheries and aquaculture are subjects of reciprocal relations with other two of its components.

Environmental degradation is caused by the actions of the main users of oceans and in-land water bodies, such as fishing and aquaculture, irrigation, tourism, maritime transport, renewable energy production, desalination of water, mining on the seabed, marine biotechnology. Thus, the desire to separate socio-economic development from this environmental degradation formed the basis of the concept Blue growth, launched by FAO in 2013.

The concept, defined as the Blue Growth initiative, is formulated as follows: "the use of natural renewable resources of the oceans, inland waters and wetlands for economic returns in a way that minimizes environmental degradation, loss of biodiversity and unsustainable use of water resources, and maximizes economic and social benefits" (FAO 2015).

This definition links the concepts of "sustainable development" and "blue bioeconomy" and acts in the framework of the implementation of sustainable development goals.

The Blue Growth initiative aligns with the need of conservation and sustainable use of the oceans, seas and marine resources for sustainable development (SDG 14) and contributes to its key objectives:

- reduce and control marine pollution (14.1);
- protect and restore ecosystems (14.2);
- reduce ocean acidification (14.3);
- regulate harvesting and end overfishing, illegal, unreported and unregulated fishing and destructive fishing practices (14.4);
- eliminate fisheries subsidies which contribute to overcapacity and overfishing (14.6);
- increase the economic benefits from sustainable use of marine resources (14.7);
- increase scientific knowledge, research and technology for ocean health (14.a);
- provide access for small-scale artisanal fishers to marine resources and markets (14.b);
- enforce international sea law concerning the sustainable use of the oceans (14.c).

The Blue Growth Initiative is an innovative, integrated and multisectoral approach to the management of aquatic resources aimed at maximizing the ecosystem goods and services obtained from the use of oceans, inland waters and wetlands, while also providing social and economic benefits (Yarkina 2019).

Consideration of FAO programs on the issue of sustainable development of fisheries and aquaculture made it possible to highlight a number of fundamental characteristics of the Blue Growth initiative that are relevant in the context of blue bioeconomy:

- the object to which it is directed is a responsible management of water resources aimed at inclusive growth, which should contribute to ensuring sustainable development in all three components (economic, social, environmental);
- focus areas of the initiative are fisheries, aquaculture, fish processing, as well as eco-system (ecological) services, trade and social protection;
- methodological approach to the implementation of the initiative is integrated and is based on the idea of integrating fisheries with other users and services of aquatic ecosystems and applying an integrated methodological approach;
- the mission of the initiative is to restore the productive potential of the oceans, seas, inland water bodies;
- the idea of the initiative is to create favorable conditions for transforming the population employed in fishery and aquaculture from a user of resources into their sustainable owner;
- the principle on which the initiative is based is the recognition of the absolute necessity of the ecological services provided by aquatic ecosystems (the air we breathe; the food we eat; the water we drink) from the point of view of human well-being.

Considering the blue bioeconomy as a factor of sustainable development of modern society through the sustainable development of the fisheries and aquaculture sector, it is necessary to emphasize the key principles, without the implementation of which the fishery sector will not be able to fulfill its own mission—these are responsibility, rationality and balance (Yarkina and Logunova 2021).

15.5 Conclusion

The blue bioeconomy as a branch of economic activity is designed to address the key economic and social problems of modern society without damage to the ecosystems of the oceans. Thus, it ensures sustainable development, the idea of which is to meet the needs of living generations, without depriving future generations of the opportunity to meet their needs as well.

The fisheries sector is directly involved in achieving sustainable development goals such as ending hunger, ensuring food security and improving nutrition; ensuring healthy lifestyles; promoting sustained, inclusive and sustainable economic growth, full and productive employment and decent work for all; ensuring the transition to sustainable consumption and production; conservation and sustainable use of the oceans, seas and marine resources for sustainable development.

The growing needs, caused primarily by the growth in the world's population due to the degradation of natural biological resources, predetermine the shift in emphasis in the development of the fishery sector in favor of aquaculture. In turn, aquaculture has the potential to fight against hunger and poverty, to ensure food security and productive employment, as well as to implement other sustainable development goals.

The existing environmental problems associated with the management of fisheries and aquaculture, climate change and the growing anthropogenic and technological pressures represent the most important group of threats to the sustainable development of the fishery sector. It predetermined the Blue Growth initiative, focused on maximizing environmental benefits and services provided by the aquatic environment while providing benefits of a social and economic nature.

For the further development of fisheries and aquaculture sector, it is necessary to expertly manage fisheries. As for aquaculture, it is essential to use methods that contribute to the sustainable use of resources and the conservation of aquatic biodiversity, i.e. act on sustainable blue bioeconomy principles.

References

Austen MC, Andersen P, Armstrong C, Döring R, Hynes S, Levrel H, Oinonen S, Ressurreição A et al (2019) Valuing marine ecosystems—taking into account the value of ecosystem benefits in the blue economy. In: Future science brief 5 of the European Marine Board. Ostend, Belgium

Bruckner M, Häyhä T, Giljum S, Maus V, Fischer G, Tramberend S, Börner J (2019) Quantifying the global cropland footprint of the European Union's non-food bioeconomy. Environ Res Lett 14 (4), art no 045011

Hasselström L, Gröndahl F (2021) Payments for nutrient uptake in the blue bioeconomy—when to be careful and when to go for it Marine. Pollut Bull 167:112321

Hasselström L, Thomas J-B, Nordström J, Cervin G, Nylund GM, Pavia H, Gröndahl F (2020) Socioeconomic prospects of a seaweed bioeconomy in Sweden. Sci Rep 10(1):1610

Lopes D, Melo T, Rey F, Calado R, Rosário Domingues M et al (2021) Insights of species-specific polar lipidome signatures of seaweeds fostering their valorization in the blue bioeconomy. Mar Pollut Bull 167:112321

Silver JJ, Gray NJ, Campbell LM, Fairbanks LW, Gruby RL (2015) Blue economy and competing discourses in international oceans governance. J Environ Dev 24(2):135–160

Sustainable development goals. https://www.un.org/sustainabledevelopment/ru/sustainable-development-goals/

The FAO Blue Growth Initiative and aquaculture (2015). https://www.fao.org/cofi/43732-0d96f652208820299a272b5bd477b0406.pdf

The State of World Fisheries and Aquaculture (2020). FAO, Rome, Italy. https://doi.org/10.4060/ca9229ru

The State of World Fisheries and Aquaculture 2018—meeting the sustainable development goals. Rome (2018). Licence: CC BY-NC-SA 3.0 IGO. FAO, Rome. https://www.fao.org/3/i9540EN/i9540en.pdf

Yarkina N (2019) Sustainable development and the Blue Growth initiative. Kerch State Maritime Technol Univ Electron J 3:117–131 (in Russian)

Yarkina N, Logunova N (2021) The concept blue growth as a way for sustainable development of the fisheries. E3S Web Conf 244:03021

Part IV
Fishing

Chapter 16 Structure of Catches and Dynamics of Fisheries in the Azov-Black Sea Basin Within Krasnodar Krai

A. V. Abramchuk, N. G. Pashinova, and G. A. Moskul

Abstract The structure and volumes of catches of commercial species of aquatic biological resources in the Azov and Black Seas within the territory of Krasnodar krai have been analyzed. Information is given about the main families and taxa, the specimens of which form the basis of the fishery. About two hundred companies carry out commercial fishing in the Black Sea and Sea of Azov within the territory of Krasnodar krai. Taxa belonging to 13 families are identified in the structure of fish catches in the Black Sea. The representatives of such families as Clupeidae, Engraulidae and Mullidae are of the highest commercial importance; they account for 97% of aquatic biological resources harvested by industrial fishing. The representatives of other 10 families account for 3% of catches. In the Sea of Azov, taxa belonging to 10 families were objects of industrial fishing. The commercial fishing is based on three families, Clupeidae, Mugilidae and Cyprinidae. In total, the share of these three families accounts for 92% of the total catches of aquatic biological resources.

Keywords Azov-Black Sea basin · Fisheries · Commercial fish · Catch · Krasnodar krai

16.1 Introduction

The Azov-Black Sea basin has a rich fishing history and currently provides the surrounding regions with fresh high-quality products. At least 184 species and subspecies of fish inhabit the Black sea, and a fifth of them are of commercial importance (Yakovlev 1995; Vasilieva 2007). In the 1970s and 1980s, the USSR alone caught about 200 thousand tons of fish and seafood in the Black Sea, while the Black Sea subspecies of anchovy (*Engraulis encrasicolus* (Linnaeus)), Black Sea sprat (*Sprattus sprattus phalericus* (Linnaeus)), whiting (*Merlangius merlangus euxinus* Nordmann), horse mackerel (*Trachurus mediterraneus ponticus* Aleev), spiny dogfish (*Squalus acanthias* Linnaeus) dominated the catches.

A. V. Abramchuk (✉) · N. G. Pashinova · G. A. Moskul
Kuban State University, Stavropol Str. 149, 350040 Krasnodarad, Russian Federation
e-mail: apilab@yandex.ru

A. G. Arkhipov (ed.), *Sustainable Fisheries and Aquaculture: Challenges and Prospects for the Blue Bioeconomy*, Environmental Science and Engineering, https://doi.org/10.1007/978-3-031-08284-9_16

There are 79 species including 47 marine, 7 anadromous, 12 semi-anadromous and 13 freshwater ones in the ichthyofauna of the Sea of Azov (Yakovlev 1995). The Sea of Azov is the most productive region of the World Ocean. At the beginning of the last century, fish catches per hectare of its water area were 7 times higher than in the Caspian, 12 times higher than in the Baltic and 35 times higher than in the Black Sea (Yakovlev 1995).

The increasing pressure of some environmental factors, against the background of intensive fishing at the end of the twentieth century, led to a significant reduction in the marine aquatic biological resources of the Azov-Black Sea basin (Yakovlev 1995). At the same time, it should be noted, that at the current stage, the general approach of fishing in the Black Sea and the Sea of Azov is based on the principle of rational use of aquatic biological resources in accordance with the state of stocks of exploited populations. The scientific research makes it possible to delicately control the sustainable fishing volumes of each particular species of aquatic biological resources.

16.2 Materials and Methods

The fishing of aquatic biological resources in the Black and Azov Seas is carried out by fishing enterprises and commercial fishermen of the southern regions of Russia (the Republic of Crimea, the city of Sevastopol, Krasnodar krai and Rostov oblast). According to the data of the Azov-Kuban Department for Fish Protection of the Azov-Black Sea Territorial Administration of the Federal Agency for Fisheries of Krasnodar krai, about two hundred enterprises and commercial fishermen (124 and 68, respectively) were engaged in industrial fishing in the Black Sea and the Sea of Azov (125 and 68, respectively).

The purpose of the study is to analyze the structure and volume of catches of commercial fish species in the Azov-Black Sea basin within the territory of Krasnodar krai to ensure sustainable fishing.

16.3 Results

In 2019, fishing enterprises of Krasnodar krai caught about 14.4 thousand tons of fish and aquatic invertebrates in water areas of the seas, including 12.6 thousand tons in the Black Sea and 1.7 thousand tons in the Sea of Azov. In 2020, the catches reduced and amounted to 11.5 and 1.6 thousand tons, respectively. In the Azov-Black Sea Fishery basin, species from 15 families of marine and freshwater fish are of commercial importance (Table 16.1).

The commercial value of these water areas in the seas is determined not only by fish resources, but also by invertebrates (mollusks, arthropods), but their catch volumes are extremely low. Moreover, the main bulk of invertebrates was harvested in the

Table 16.1 Fisheries of aquatic biological resources in the Black and Azov seas within Krasnodar krai in 2019–2020

Species of aquatic biological resources	Catch volume, t			
	Black Sea		Sea of Azov	
	2019	2020	2019	2020
1. Family Squalidae				
Spiny dogfish (*Squalus acanthias* Linnaeus)	11.5	10.1	0.0	0.0
2. Family Dasyatidae				
Common stingray (*Dasyatis pastinaca* Linnaeus)	49.3	47.5	0.0	0.0
3. Family Clupeidae				
Black Sea shad (*Alosa immaculata* (Bennett))	14.1	13.0	2.3	2.4
Black Sea sprat (*Clupeonella cultriventris* (Nordmann))	0.0	0.0	420.0	375.7
Sprat (*Sprattus sprattus phalericus* (Linnaeus))	3299.5	3022.8	0.0	0.0
4. Family Engraulidae				
European anchovy (*Engraulis encrasicolus* (Linnaeus))	8350.4	6988.2	18.1	12.9
5. Family Scophthalmidae				
Black Sea turbot (*Psetta maeotica* (Pall.))	60.9	56.2	5.1	6.6
6. Family Gadidae				
Whiting (*Merlangius merlangus euxinus* Nordmann)	0.2	0.2	0.0	0.0
7. Family Mullidae				
Red mullet (*Mullus barbatus ponticus* Essipov)	536.2	1155.7	10.3	16.8
8. Family Carangidae				
Horse mackerel (*Trachurus mediterraneus ponticus* Aleev)	227.5	117.2	2.2	2.3
9. Family Mugilidae				
So-ioy mullet (*Liza haematocheilus* (Temminck and Schlegel))	0.2	1.4	638.9	640.0
Golden gray mullet (*Liza aurata* (Risso), flathead gray mullet (*Mugil cephalus* Linnaeus)	43.8	47.6	30.8	70.3
10. Family Atherinidae				
Big-scale sand smelt (*Atherina mochon pontica* Eichwald)	0.1	0.1	3.7	2.3
11. Family Pomatomidae				
Bluefish (*Pomatomus saltatrix* (Linnaeus))	0.2	0.2	0.0	0.0
12. Family Sparidae				
Picarel (*Spicara smaris* (L.))	21.0	20.1	0.0	0.0
13. Family Belonidae				
Garfish (*Belone belone euxini* (Günther))	14.6	15.3	0.3	0.2

(continued)

Table 16.1 (continued)

Species of aquatic biological resources	Catch volume, t			
	Black Sea		Sea of Azov	
	2019	2020	2019	2020
14. Family Cyprinidae				
Roach (*Rutilus heckelii* (Nordmann))	0.0	0.0	15.3	17.1
Bream (*Abramis brama* (Linnaeus))	0.0	0.0	8.9	9.3
Crucian carp (*Carassius sp.*)	0.0	0.0	496.8	368.3
15. Family Gobiidae				
Goby (*Neogobius sp.*)	0.0	0.0	7.2	8.4
Other fish species				
Other freshwater	0.0	0.0	52.2	34.1
Other marine	14.8	14.4	0.0	0.0
Commercial invertebrate species				
Rapa whelk (*Rapana thomasiana thomasiana* Crosse)	0.1	0.1	0.3	0.1
Shrimp (*Palaemon sp.*)	0.3	0.4	7.9	16.2
Brine shrimp (*Artemia salina* (Linnaeus))	0.0	0.0	4.2	5.4
Total	**12,644.7**	**11,510.5**	**1724.5**	**1590.4**

Source "Compiled by the authors" according to the data of the Azov-Kuban Department of State Control, Supervision and Protection of Aquatic Biological Resources of the Azov-Black Sea Territorial Administration of the Federal Agency for Fisheries"

water area of the Sea of Azov. Currently, the bivalve mollusk fisheries have certain limitations, primarily related to the implementation of programs for the restoration of natural populations.

The taxa belonging to 13 families are identified in the structure of fish catches in the Black Sea. The representatives of such families as Clupeidae, Engraulidae, and Mullidae are of the highest commercial importance; they account for 97% of aquatic biological resources harvested as a result of commercial fishery. The representatives of other 10 families account for about 3% of catches (see Fig. 16.1).

It should be mention, that representatives of the families Cyprinidae and Gobiidae were nor recorded in the catches during the period under consideration.

The maximum catches both in 2019 (12.4 thousand tons) and in 2020 (11.3 thousand tons) were composed of such species as European anchovy, *Engraulis encrasicolus* (8.4 and 7.0 tons), sprat, *Sprattus sprattus phalericus* (3.3 and 3.0 thousand tons), red mullet, *Mullus barbatus ponticus* (0.5 and 1.2 thousand tons), and horse mackerel, *Trachurus mediterraneus ponticus* (0.2 and 0.1 thousand tons).

In the Sea of Azov, the objects of commercial fishery were taxa belonging to 10 families. The fishery is based on three families: Clupeidae, Mugilidae and Cyprinidae. In total, these three families account for 92% of the total catches of aquatic biological resources (see Fig. 16.2).

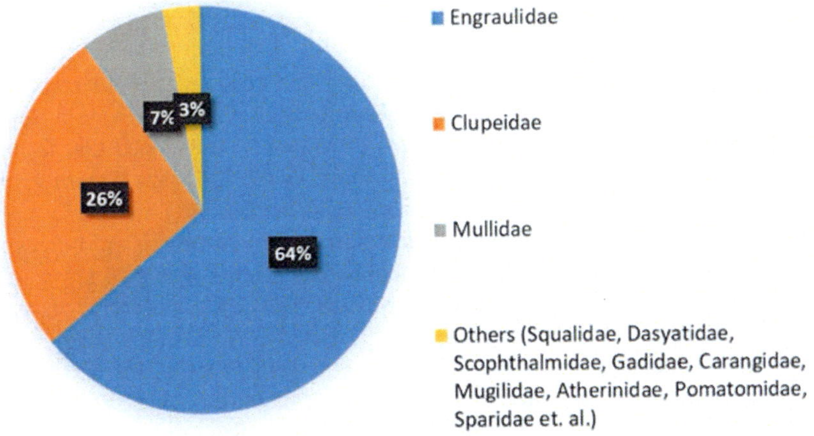

Fig. 16.1 Proportion of representatives of particular fish families in catches in the Black Sea within Krasnodar krai. Compiled by the authors

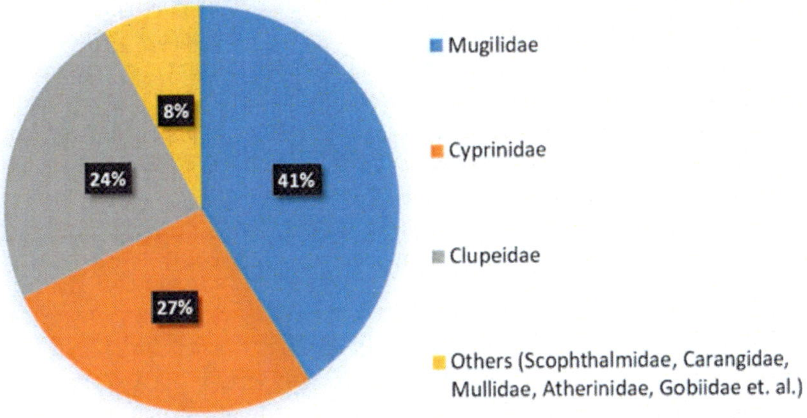

Fig. 16.2 Proportion of representatives of particular fish species in catches in the Sea of Azov within Krasnodar Krai. Compiled by the authors

In taxonomical terms, the catches on average for the two analyzed years are composed of *Clupeonella cultriventris* (0.4 thousand tons), *Liza haematocheilus* (0.6 thousand tons) and *Carassius sp.* (0.5 thousand tons).

16.4 Discussion

The water areas of the southern seas of Russia within the territory of Krasnodar krai have a rich history of fishing industry and currently continue to provide surrounding regions with fresh high-quality products. It should be mentioned that in modern conditions, European anchovy (*Engraulis encrasicolus* L.), sprat (*Sprattus sprattus phalericus* (L.)), red mullet (*Mullus adult ponticus* Essipov) and horse mackerel (*Trachurus mediterraneus ponticus* Aleev) are important fisheries resources in the Black Sea. These species dominate in commercial catches (Table 16.1).

In taxonomical terms, the catches in the Sea of Azov within the territory of Krasnodar krai are basically represented by Black Sea sprat (*Clupeonella cultriventris* (Nordmann)), so-iuy mullet (*Liza haematocheilus* (Temminck and Schlegel)) and crucian carps (*Carassius sp.*) (Table 16.1). It should be noted that in past years, valuable fish species such as sturgeon, vimba bream, Danube bleak, zander, roach dominated in catches. At present, many of them have lost their commercial significance. Reduced catches of valuable fish species were reported by other authors (Volovik et al. 1992; Balykin and Kutsyn 2017; Makarov et al. 2000; Chebanov et al. 2004 etc.). Intensive economic activity in the region (regulation of rivers and unavailability of natural spawning grounds, significant commercial pressure, illegal fishing, decrease in freshwater runoff, etc.) was the reason for the change in the sustainable catch of these species of aquatic biological resources. In order to ensure sustainable fishing of aquatic biological resources in the waters of the Azov and Black Seas, based on up-to-date information about the state of populations of commercial species, scientific institutions conduct studies aimed at determining the total allowable catches, time and fishing regulations.

16.5 Conclusion

Over recent years, the commercial ichthyofauna of the Black and Azov Seas has been undergoing significant qualitative and quantitative changes, in our opinion, primarily due to the intensity of fishing in the states of the Azov-Black Sea basin and adverse effect of some abiotic and biotic environmental factors (decrease in the water level, introduction of alien species, eutrophication, etc.). It is apparent that in order to normalize the ecological situation and to increase the volume of catches in the Sea of Azov and the Black Seas coordinated nature conservation and scientifically-based fishing activities are necessary not only at the national but also at the international level.

References

Balykin PA, Kutsyn DN (2017) Modern salinization of the Sea of Azov and prospects of fisheries. In: All-Russian scientific conference “Aquaculture: world experience and Russian developments”. Southern Scientific Center of the Russian Academy of Sciences, Rostov-on-Don, pp 388–390 (in Russian)

Chebanov MS, Galich EV, Chmyr YuN (2004) Guidelines for breeding and rearing of sturgeon. Rosinformagrotech, Moscow (in Russian)

Makarov EV, Zhiteneva LD, Abrosimova NA (2000) Living fossils are close to extinction: a scientific essay on sturgeon. DSTU, Rostov-on-Don (in Russian)

Vasilieva ED (2007). Fishes of the Black Sea. VNIRO, Moscow (in Russian)

Volovik SP, Makarov EV, Semenov AD (1992) The state of the ecosystem and fish stocks of the Sea of Azov, measures for their protection. In: The 2nd international conference “Problems of studying and rational use of biological resources of the marginal and inland seas of the CIS”. Southern Scientific Center of the Russian Academy of Sciences Rostov-on-Don, pp 23–28 (in Russian)

Yakovlev VN (1995) The state of biological resources of the Black and Azov Seas. YugNIRO, Kerch (in Russian)

Chapter 17
Impact of Gidrometerogical Conditions on Smelt Spawning Migration and Catch Fluctuations in the Rivers of Curonian Lagoon Basin

Anna Burbakh and Sergey Shibaev

Abstract Smelt together with dwarf smelt (resident form) are the most important fishing targets in Kaliningrad oblast in the spring. At the beginning of the twentieth century, their catches reached up to 9 thousand tons (Germany—USSR); currently, the catches do not exceed a thousand tons. Until 2010, the fishing was performed directly in the Curonian Lagoon, since 2011 the role of the Curonian Lagoon-Neman River transit system in fishery statistics has been increasing. The transit system includes the lower course of the Neman River (the Skirvite arm, the Vitine and Atmata rivers) and the Matrosovka River. The restructuring of the fishery system was affected by climatic factors and changes in the legislative regulation of smelt fisheries. The climatic factors, first, the ice regime, determined the new dislocation of the fishery. Changes in fishery legislation in 2011 were accompanied by engagement of a larger number of fishing enterprises and fishermen in smelt fishing and an increase in the fishing efficiency. At the same time, the problem of accurately predicting the allowable volume of catches still exists, because smelt is a species that is very susceptible to the environmental impact. We analyzed the dynamics of spawning migration and catches depending on hydrometeorological factors, such as water temperature during different periods of spawning migration, the rate of water warming, the day of fish migration to the rivers (length of daylight), wind regime. According to the data obtained, it can be concluded that the migration pattern and the volume of catches depend on the rate of water heating and photoperiod, the earlier the migration starts and more gradually water heating occurs, the higher the total catch will be.

Keywords Smelt · Dwarf smelt · Curonian Lagoon · Neman River · Catch · Water temperature · Migration duration · Wind regime

---

A. Burbakh (✉) · S. Shibaev
Kaliningrad State Technical University, Sovetsky avenue, 1, 236022 Kaliningrad, Russia
e-mail: anna.burbakh@klgtu.ru

S. Shibaev
e-mail: shibaev@klgtu.ru

A. G. Arkhipov (ed.), *Sustainable Fisheries and Aquaculture: Challenges and Prospects for the Blue Bioeconomy*, Environmental Science and Engineering, https://doi.org/10.1007/978-3-031-08284-9_17

17.1 Introduction

The smelt is an anadromous species that is very susceptible to environmental effects (Repečka and Gerulaitis 1994). Numerous studies of Russian and foreign researchers have been devoted to the smelt biology as an important commercial species (Gaigalas 1980; Manyukas 1959; Shibaev et al. 2012). Despite the more or less studied biology of the species, the problem of accurately predicting the allowable volume of catches still exists (Ryabchun et al. 2020), i.e. it can be concluded that the smelt fishery is affected rather not by the population size, but by hydrometeorological factors that can affect the spawning migration of smelt, and, accordingly, its yield. This issue was considered only in the work of the Lithuanian researcher in regard to the catches in the Curonian Lagoon (Svagždys 2009).

Table 17.1. Both smelt (*Osmerus eperlanus eperlanus* L.) which lives in the Baltic Sea, and its resident freshwater form (*Osmerus eperlanus eperlanus morpha spirinchus* L.) play an import role in commercial fisheries; they together belong to the family Osmeridae and in the fishery statics, their total catches are usually reported. Fishery is conducted on spawning routes, namely in the Curonian Lagoon and in the Curonian Lagoon-Neman River transit zone, which includes the lower course of the Neman River (the Skirvite arm, the Vitine and Atmata rivers) and the Matrosovka River. The fishery in the transit zone was not sufficiently developed until 2011 for several reasons. First, the fisheries regulation based on the system of the total allowable catch did not allow a large number of commercial enterprises and fishermen to perform fishing, since the quota was distributed among four fishing collective farms according to the historical principle. This approach often led to an underutilization of the quota, especially in years when the ice breakup in the lagoon occurs later, and it was impossible to use set gillnets in the lagoon for the

Table 17.1 Temperature regime during the spawning migration of smelt, 2012–2020 (prepared by the authors)

Year	Migration start date	Duration, days	Water temperature at the start of migration, °C	Water temperature during high speed of migration, °C	Water temperature at the end of migration, °C
2012	20 March	8	0.7	3.8	3.8
2013	6 April	13	1.4	3.5	3.8
2014	11 March	15	1.0	3.9	4.3
2015	27 February	22	1.0	2.0	2.1
2016	27 February	28	2.0	3.0	3.4
2017	03 March	23	0.3	3.0	3.8
2018	25 March	24	1.0	3.0	5.9
2019	03 March	20	2.0	4.0	5.8
2020	27 February	28	2.0	3.2	4.2

quota utilization, and therefore smelt migrated to rivers where commercial fishermen were ready to fish, but their quotas were insufficient for large volumes of catches. Second, the lack of a sufficient number of fishing sites on the rivers, since after the collapse of the USSR, the main fishing sites on the rivers of the Curonian Lagoon basin became part the Lithuanian territory. Russian fishermen were traditionally fishing in the lower reaches of the river, mainly in the Skirvite arm. But in 2012, the possible recommended catch (RC/PC) became an alternative to the system of fishery regulation based on total allowable catch (TAC) which made it possible to develop fisheries in rivers (Anuryeva et al. 2015; Burbakh et al. 2019; Burbach 2021).

The aim of the work is to identify the main hydrometeorological factors that affect the dynamics of the spawning migration and smelt fisheries in the Curonian Lagoon-the Neman River transit system.

17.2 Materials and Methods

This work is based on the data of monitoring surveys at fishing sites in the period from 2012 to 2020. Fish sampling was conducted in different sections of the Curonian Lagoon-Neman River transit system, depending on the tasks: in the lower reaches of the Neman River, in the Matrosovka River outflowing the Neman River downstream of the city of Sovetsk, and in the Deyma River (2020) near the fishing sites of fishing brigades in the current year (see Fig. 17.1).

All data on temperature and water level fluctuations in the transit system were obtained directly by the authors.

17.3 Results

The maximum historical catch of smelt was 8.8 thousand tons in 1935 (Osadchiy 2000), in the modern period such a value has never been achieved. The Russian catches after the collapse of the USSR range from 24 to 541 tons, averaging 188 tons.

The year of 2010 is a turning point in the history of smelt fishery. That year, the winter was very cold and prolonged, the lagoon was covered with ice for a long time, therefore, fishing with set gillnets was impossible. As a result, the total catch in 2010 amounted to 23 tons; it is the minimum value for the last 30 years.

Since 2011, there was a tendency in Russia to relocate fisheries to the rivers of the Curonian Lagoon basin, which has become possible for several reasons. First, in some years the ice breakup starts late in the lagoon. Second, the fishing regulations changed. Thus, in 2012, smelt was transferred to the category of aquatic biological resources, for which the total allowable catch (TAC) was not established, but the recommended (possible) catch (RC/PC) was defined. As a result of the transition to the system of the recommended catch (RC/PC), all interested fishing enterprises

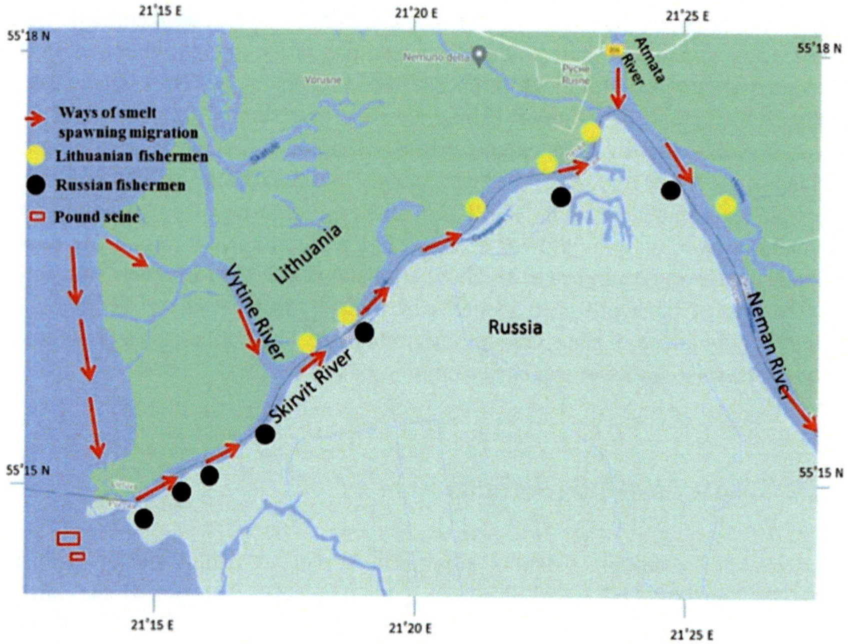

Fig. 17.1 Map of the fishing brigades locations and smelt migration ways (compiled by the authors)

and fishermen receive fishing permits for the use of aquatic biological resources, and fishing is carried out according to the "Olympic system". The new regulatory system has made it possible to increase the efficiency of fishing. The allowable volume of catches increased from 38 to 81% (Fig. 17.2).

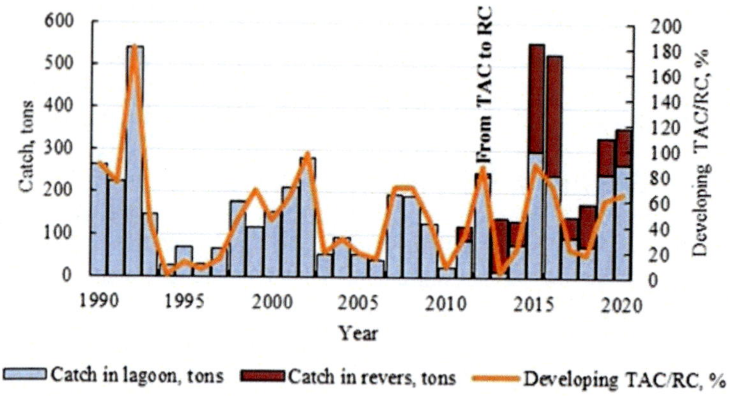

Fig. 17.2 Long-term dynamics of commercial catches of smelt and developing of the allowed volume of production

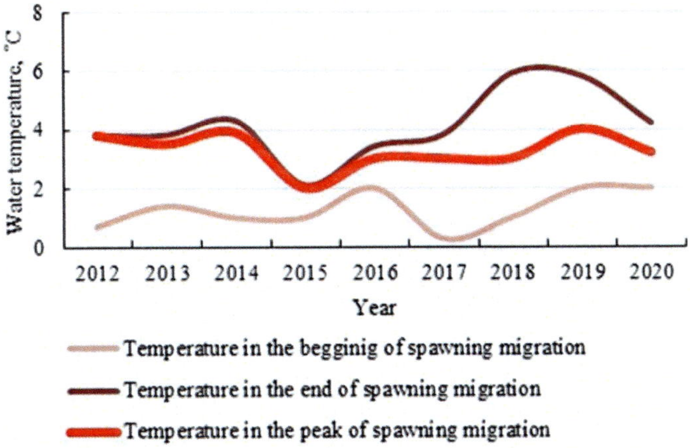

Fig. 17.3 Temperature regime during spawning migration of smelt to the rivers of the Curonian Lagoon basin (compiled by the authors)

In some years, the catch volume exceeds the recommended one, but this fact does not have a negative impact on the smelt stock.

We paid attention to the main hydrological factors that can affect the dynamics of the spawning migration and catches such as water temperature during different periods of spawning migration, rate of water warming, the day fish enter the rivers (the length of daylight), wind regime.

During the research period, smelt enters the Curonian Lagoon-Neman River transit zone in different periods, usually at the end of February or early March, less often at the end of March or mid-April, usually at a temperature of 1.0–1.5 °C, only in 2016 it was recorded at a temperature of 2 °C. The spawning migration ends at a water temperature of about 4–6 °C, i.e. at spawning temperatures, the period of intensive migration is observed at a temperature of 3–4 °C (see Table 17.1, Fig. 17.3).

When comparing the data on catches in the rivers with the average water temperature during the period of intensive migration, when the maximum catches are recorded, it is found that the temperature affects rather not the amount of total catch and the duration of migration, but the time of entry and termination of migration, this in turn determines the results of fishing. Thus, in 2012, when the average temperature was 3.5 °C during the period of intensive migration, the yield in the rivers was 252 tons, and in 2014, at the same temperature, the yield was 54 tons. At an average temperature of 3 °C during the intensive migration, the catches differ, amounting to 275 and 50 tons in 2016 and 2017, respectively (see Fig. 17.4).

Another important factor determining spawning migration is the rate of water warming. At slow water warming, the spawning migration is gradual and long. Accordingly, the catches per effort are low, but the total catch is higher than in years when warming period is short. The most durable migrations were recorded in 2015 and 2016 (23 and 30 days, respectively); at that time the catches in the rivers were maximal and amounted to 257 (2015) and 286 tons (2016) (see Fig. 17.5).

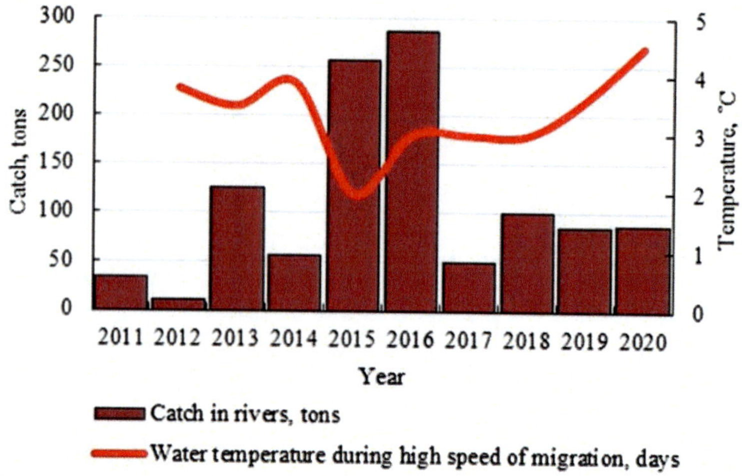

Fig. 17.4 Temperature regime during spawning migration and its impact on smelt catches in rivers (compiled by the authors)

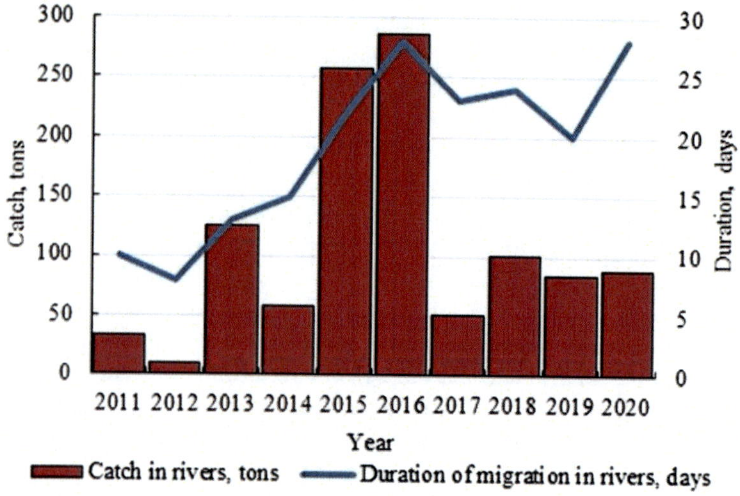

Fig. 17.5 Relationship between smelt catches in rivers and duration of spawning migration (compiled by the authors)

At the same time, the rate of warming has an impact in combination with other hydrometeorological factors, which does not allow conducting the direct correlation analysis. Thus, in 2018, the duration of the spawning migration was 22 days, and the yield was low (100 tons).

The study of the temporal dynamics of the spawning migration showed that the date of its beginning is not always correlated with the water temperature, but has its own deterministic period. It was found that at an earlier migration into rivers, the yield is higher (the correlation coefficient is 0.7, which indicates a strong dependence). Conversely, the later the water temperature reaches the optimum for the beginning of the spawning migration, the smaller the catch value is (see Figs. 17.6 and 17.7).

The water level during the spawning migration varied between 1.9 and 3.4 m. The average fluctuation of the level during the spawning migration of smelt in the Skirvite River is 10–15 cm. The dependence of the water level on the daily dynamics of catches was not found (see Fig. 17.8). The effect of the water level on the total catch in the rivers was not also found (see Fig. 17.9).

The wind regime is an important factor during the spawning migration. It should be mentioned that, an important condition for smelt to enter the Neman River is "silence", i.e. when the ice breaks or the wind is strong, the fish cannot enter the rivers because they are deterrent factors. The depth of the Skirvite River in the

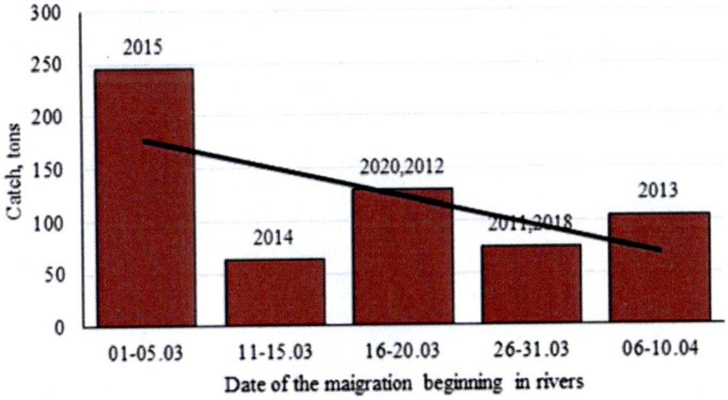

Fig. 17.6 Relationship between smelt catch in the rivers and the day of entry into the rivers of the Curonian Lagoon basin by years (compiled by the authors)

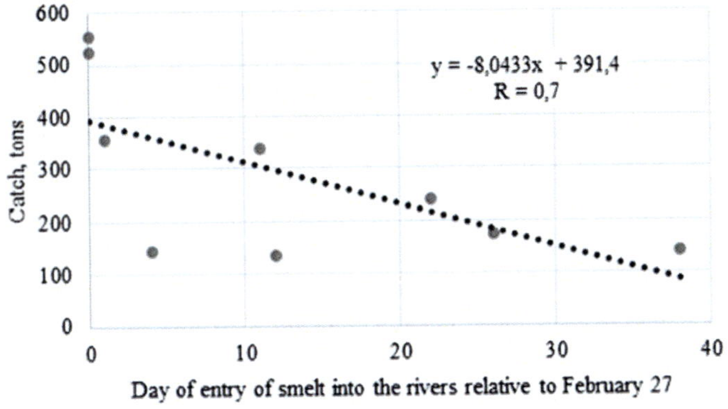

Fig. 17.7 Relationship between the catch of smelt in the rivers of the Curonian Lagoon basin and the day of entry into the rivers relative to the date of the earliest start of spawning migration (compiled by the authors) (x-day of entry of smelt into the rivers relative to February 27, y-catch)

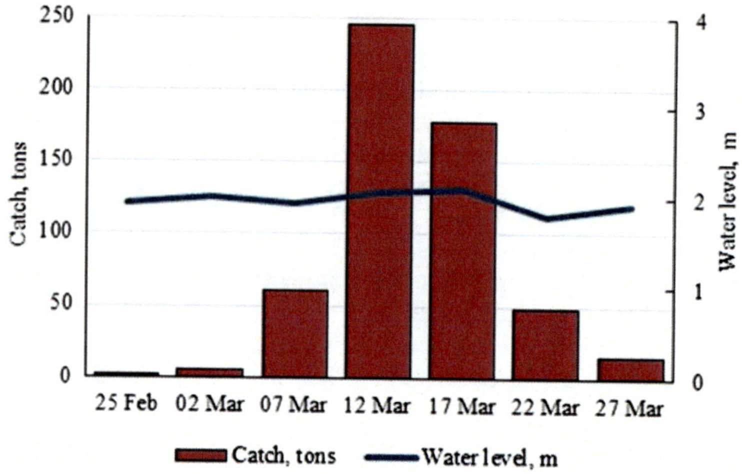

Fig. 17.8 Level regime during the spawning migration period and its impact on the smelt catch by the dates of fishing (compiled by the authors)

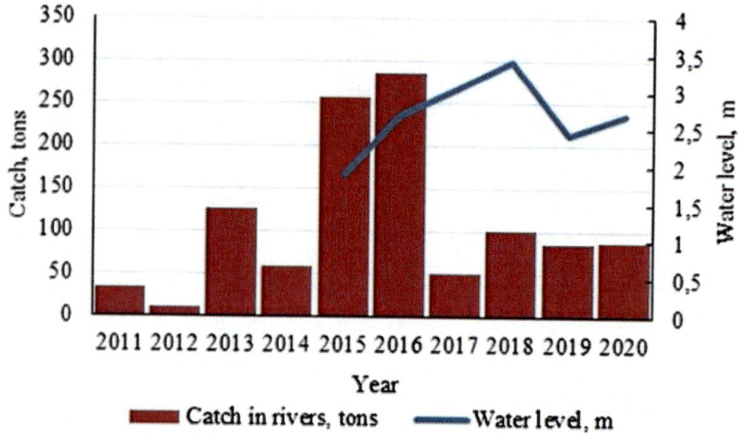

Fig. 17.9 Level regime during spawning migration and its impact on smelt catches in rivers (compiled by the authors)

mouth part during this period is no more than 0.5 m, which makes it impossible for the smelt to descend to avoid the action of the wind. Provided that strong winds blow during the fishing season, the efficiency of fishing also decreases, due to the fact that the "shoal breaks" and the fish move separately.

17.4 Conclusion

In modern conditions, there has been a redistribution of smelt catches between the Curonian Lagoon and the rivers of its basin. In some years, the share of smelt catches in the rivers reaches 90% (2013), on average over the past 10 years, the share of catches in the rivers has been 40–50%. The period of smelt spawning migration in the rivers is at the end of February-beginning of April, and depends on hydrological factors. The beginning of migration is usually timed to water warming to a water temperature of 1.0–1.5 °C, the peak of the migration is observed at a temperature of 3–4 °C, at a higher temperature, solitary specimens are found in catches. The slow water warming and the early beginning of the migration determine the maximum catches of smelt, so in 2014 the spawning migration lasted 15 days and the total catch was 58 tons, whereas in 2016, the catch was 5 times higher for 28 days.

Due to the different fishing modes, set gillnets are used in the lagoon, and beach seine in the rivers, a separate system of fisheries regulation is necessary. The transition from the regulation of smelt fisheries according to the system of the total allowable catch, which is distributed by quotas among fishing enterprises to fisheries according to the "Olympic system", the utilization of the total quota by all fishing enterprises, ensured self-regulation of fishing depending on the intensity of the spawning migration in each particular year.

Due to the fact that intensive spawning migration and smelt fishery lasts 5–7 days, and then stops, the by-catch of related species, with the exception of ruff, corresponds to the concept of an ecosystem approach.

References

Anuryeva AS, Sokolov AV, Shibaev SV, Luneva EV (2015) Migration of smelt (*Osmerus eperlanus eperlanus* L.) in the Neman River in the zone of possible influence of the Baltic Nuclear Power Plant. Conference materials: Environmental safety of nuclear power plants: II scientific and practical conference with international participation, dedicated to the 70th anniversary of the nuclear industry in Russia. Axios, Kaliningrad (in Russian)

Burbach AS (2021) Characteristics of the spawning part of the population of the European smelt (*Osmerus eperlanus eperlanus* L.) in the transit system Curonian Lagoon—the Neman River (abstract of Ph.D. thesis in biological sciences). Kaliningrad State Technical University, Kaliningrad (in Russian)

Burbakh AS, Shibaev SV, Sokolov AV (2019) Evolution of the fishery for smelt (Osmerus eperlanus eperlanus L.) and smelt (Osmerus eperlanus eperlanus morpha spirinhus L.) in the rivers of the Curonian Lagoon of the Baltic Sea Basin. Fisheries 3:85–89 (in Russian)

Gaigalas KS, Mishtautite VT (1980) The main features of ecology and dynamics of commercial catches of the anadromous form of the European smelt *Osmerus eperlanus* (L.) in the Curonian Mares Bay and the lower reaches of the river Nemunas. Quest Ichthyol 20(4):625–634

Manyukas (1959) Ichthyofauna, state of stocks and fisheries in the Kursyumares Bay. Kurshumares. Results of comprehensive research, Vilnius, 403 pp

Osadchiy VM (2000) Regulation of fisheries and the strategy for the use of fish resources in the Curonian Lagoon (abstract Ph.D. thesis in biological sciences). Kaliningrad State Technical University, Kaliningrad (in Russian)

Repečka R, Gerulaitis A (1994) Monitoring of abundance and biological characteristic of migratory fish species (Salmon, Sea trout, Vimba and Smelt) in the beginning of spawing migration. Summary. Žuvininkystė Lietuvoje 1:47–48

Ryabchun VA, Podgorny TA, Golubkova TA (2020) Current state of the European smelt *Osmerus eperlanus* (L.) population in the Curonian Lagoon and the use of adaptive models to predict the recommended catch. Proc AtlantNIRO 4(9):94–115 (in Russian)

Shibaev SV, Sokolov AV, Tylik KV, Svirina LV, Smirnova MA, Shkitsky VA (2012) Spawning run of smelt (*Osmerus eperlanus eperlanus* L.) in the rivers of the Curonian Lagoon basin. Bull Baltic Fed Univ I Kant 7:91–96 (in Russian)

Svagždys A (2009) Impact of environmental conditions on smelt catch fluctuations in the Nemunas River and the Curonian Lagoon. Ekologija 55(3–4):204–214

Chapter 18
Characteristics of the Ichthyocenosis in the Wintering Pits in the Context of the Zoogeographic Origin on the Example of Shallow Waters of the Volgograd Reservoir Basin

Irina A. Kiyashko and Vladimir V. Kiyashko

Abstract The Volgograd reservoir is a complex hydrological system, in which locations of the wintering pits have long been determined in its the deep-water channel part. Wintering pits can also be found in the floodplain areas, yet their relatively small size does not always allow determination their characteristics by the hydrometric and the bathymetric indicators. The aim of this article is to investigate the species composition of the ichthyocenosis in shallow waters of the Volgograd Reservoir basin, to determine the behavior pattern of fish populations and their involvement in the formation of winter fish aggregations in the context of the zoogeographic origin in order to use the obtained findings in fisheries. Studies were conducted in the channel wintering pits in shallow waters of the Volgograd Reservoir on the example of the Krasnoyarsk floodplain in winter 2019–2020. The floodplain is composed of not only large and small bays deeply protruding into the land, but also of flooded creeks, backwaters, lakes, small rivers and «eriks» (small channels) with preserved flow paths and a wide variety of bio-topes, which provide conditions for feeding, reproduction and wintering of a significant number of fish species. In the wintering pits of the Krasnoyarsk floodplain, were identified 18 fish species from 5 orders and 7 families, which can be divided into 6 faunistic complexes with regard to their morphological and ecological characters. Two of these, the Ponto-Caspian freshwater and Boreal plain, formed the basis of ichthyofauna (77%) of the winter aggregations. At the same time, fish aggregations from the Ponto-Caspian freshwater and the Tertiary plain freshwater faunal complexes, which include such valuable fish species as carp, bream, catfish, silver bream, and asp, in the autumn–winter period can be considered as an indicator of wintering pits.

I. A. Kiyashko (✉)
Saratov State Technical University Yu.A. Gagarina, Saratov, Russia
e-mail: coba80@mail.ru

V. V. Kiyashko
Saratov Branch of FGBNU VNIRO, Saratov, Russia
e-mail: kiyashko_81@list.ru

A. G. Arkhipov (ed.), *Sustainable Fisheries and Aquaculture: Challenges and Prospects for the Blue Bioeconomy*, Environmental Science and Engineering, https://doi.org/10.1007/978-3-031-08284-9_18

Keywords Wintering pits · Volgograd Reservoir · Species behavior · Faunistic complexes

18.1 Introduction

Rational use of natural resources of inland water bodies, increase in its productivity, and increased qualitative and quantitative characteristics of fisheries yields are one of the main tasks posed by the Blue Economy.

The commercial fishery resources in water bodies of the temperate zone of the Europe are formed during a warm period which is characterized by high water temperatures, the availability of a sufficient amount of food and favorable conditions for reproduction and feeding for most commercial fish species (Shashulovsky and Mosiyash 2010).

Winter is a special period for aquatic life. The ice cover hinders water saturation with atmospheric oxygen. The snow cover reduces the access of sunlight to the under-ice space. Feeding capacity of the water body sharply decreases. Water temperature decreases, and most of the cold-blooded animals (especially fish) become inactive. Most fish survive the winter due to the nutrients stored in summer, but the nutrient expenditure causes a decrease in its individual weight.

Previous studies conducted in inland water bodies of various types showed that wintering pits used by fish can help them to survive during unfavorable wintering conditions. Wintering pits are natural places of mass wintering of fish in rivers, lakes and reservoirs. Winter aggregations include fishes of one or several species. In flowing water bodies, wintering pits are located in the deepest places with a slow current. In lakes and reservoirs, they are typically located in the mouth part of inflowing rivers and streams, and at the outlet of underwater sources (Shashulovsky and Ermolin 2005a).

Such places are similar not so much in depth but in comfortable conditions—oxygen content, slow water velocity, high temperature, hard bottom sediments, and the lack of vegetation.

In addition, wintering pits can be located at the outlet of an underwater source in the form of a spring or an underwater river with a weak water flow. Wintering pits can be also located in a deep channel near the riverbed.

Large aggregations in wintering pits are typical to Acipenseridae (sturgeon, sterlet, etc.) and many of Cyprinidae fish (carp, crucian carp, tench, bream, etc.). The main wintering pits in large rivers were identified during the period from the beginning to the middle of the last century.

It should be noted that not all river «pits» become wintering ones, it should have a bottom area with a slow circular current. In such areas, oxygen supply for fish is sufficient, and it does not expend much of energy to resist the current.

Initially, it was believed that fish accumulate in such areas only in winter, when their activity is decreasing; they completely stop feeding or significantly reduce food consumption, and the biological role of channel pits is reduced to the wintering area of the river. However, at present, modern hydroacoustic systems detect aggregation of fish of various age groups in such areas all year around. For example, in spring, significant aggregations of fish are formed in the Kondinskaya wintering pit, and their number varies over days. When the water temperature in the Konda River exceeds 4 °C, the fish migratory activity significantly increases both upstream and downstream in the system “channel pit–tributary of the main river”, yet a significant part of fish stays in the Konda River (Chemagin 2016).

One of the main reasons for fish aggregation in such channel areas of the river is a “hydrodynamic shadow” caused by sharp changes in its bottom relief and, accordingly, nonuniform currents. Individual fish species and their aggregations are least susceptible to current effects, which helps them to stay in the occupied spatial areas of the riverbed with minimal energy consumption (Chemagin 2016).

The location of the wintering pits is determined empirically by the presence of mass fish aggregations. Over the years, the location of the pits can change, which is associated with a change in the relief. That is why it is difficult to identify the exact location of the wintering pit, and only the area of its location is therefore indicated.

In the rivers of the basin of the Volgograd reservoir, wintering pits play a special role due to long fish migration routes and the seasonal fish suffocation. In such biotopes, valuable fish species (like pike perch) survive the winter. In addition, in some pits, the commercial valuable species of fish (catfish, bream, pike, etc.) and burbot accumulate for wintering. Therefore, wintering grounds for fish should be considered as an important condition for preservation of the potential of the biological resources of water bodies in the Saratov region.

Numerous studies investigated the ichthyofauna composition in shallow water bodies of different types. Zakora (1974), Nebolsina (1974), and Cherepanov (1995) performed detail studies of shallow waters of the Volgograd Reservoir. Results of their investigations consider role and significance of the shallow waters for the reproduction of fish stocks. In addition, Zakora (1974) profoundly investigated their hydrology, bottom grounds, food resources, and ichthyofauna; and also the measures are proposed to preserve and increase fish productivity in shallow water areas. The studies of these authors has a strictly practical nature and apparently important; however, they did not sufficiently cover the issues of the state of wintering pits.

Previously, it was believed that fish lay on the pit bottom and fall into suspended anabiosis; however, subsequent observations showed that, despite of crowded, fish keep slowly moving, and the movement usually proceeds from the center of the aggregation to its edge (and vice versa). Apparently, this behavior pattern provides equal conditions for all individuals in the shoal. At the same time, some species keep feeding quite actively, for example, bream and ide. In this case, the entire shoal can change its location when feeding grounds are depleted.

In winter, fish obey natural instincts and show specific behavior patterns. To understand the process (behavior patterns), characteristics of ichthyocenosis in shallow waters of the Volgograd Reservoir should be considered in the context of the zoogeographic (faunal) origin.

18.2 Materials and Methods

The aim of the study was to investigate the species composition of ichthyocenosis in shallow waters of the Volgograd Reservoir basin, to determine the behavior pattern of fish species and their role in formation of winter aggregations in the context of the zoogeographic origin in order to use the obtained information by fisheries.

Ichthyofauna of the wintering pits was investigated in channel pits in the Krasnoyarsk floodplain of the Volgograd Reservoir in winter in 2019–2020.

The species composition of ichthyocenosis was determined on the basis of control catches and catches by anglers. In total, 12 nettings were carried out, and 78 anglers were interviewed.

The relative number of fish in catches was determined according to the scale proposed by Ermolin (2010).

The ichthyological material was processed according to the common methods (Pravdin 1966).

The study employed materials obtained by the authors and literature data on characteristics of the floodplain as a fish habitat (Nebolsina 1974).

The fish belonging to faunal complexes is provided in accordance with the studies by Nikolsky (1951), and with clarification by Nebolsina (1980).

18.3 Results

The Krasnoyarsk floodplain stretches along the left bank of the Volga River from the city of Marx to the city of Engels. It includes not only large and small bays deeply protruding into the land, but also flooded river creeks, backwaters, lakes, small rivers and «eriks» with a preserved flow and a wide variety of biotopes.

The floodplain area is about 20 thousand hectares. Along the longitudinal transect of the reservoir, it is 26 km long and 4–10 km wide. Its depths range from 0.1 to 10 m. Large and small islands separating water areas are covered with meadow vegetation, which is a substrate for fish spawning. The biotope diversity provides conditions for feeding and reproduction of a significant number of fish species.

The water level regime and the flow rate depend on the operating mode of the hydroelectric power stations (Saratov and Volgograd).

The content of dissolved oxygen and carbon dioxide, and water pH vary significantly throughout the year with the pronounced seasonal pattern. The greatest oxygen deficiency (49% saturation) can be observed in January–March.

Since aquatic life normally develops in water with a neutral or slightly alkaline pH, the study area provides a favorable habitat for aquatic organisms in terms of pH values.

In the wintering pits of the Krasnoyarsk floodplain, according the control catches and anglers catches form this study, identified 18 fish species from 5 orders and 7 families. The most numerous family is Cyprinidae (12 species). The other 6 families include only 1 species in each (see Table 18.1).

Table 18.1 Species composition of fish in the wintering pits

Species and its taxonomic position	RN
Order I. Salmoniformes	
Family 1. Esocidae Cuvier, 1816	
1. *Esox lucius* Linnaeus, 1758	+++++
Order II. Cypriniformes	
Family 2. Cyprinidae Bonaparte, 1832	
2. *Abramis ballerus* (Linnaeus, 1758)	+++
3. *A. brama* (Linnaeus, 1758)	+++++
4. *Aspius aspius* (Linnaeus, 1758)	+++
5. *Blicca bjoerkna* (Linnaeus, 1758)	+++++
6. *Carassius auratus gibelio* (Bloch, 1782)	+++++
7. *Cyprinus carpio* Linnaeus, 1759	+++
8. *Leucaspius delineatus* (Heckel, 1843)	++
9. *Leuciscus cephalus* (Linnaeus, 1758)	++++
10. *L. idus* (Linnaeus, 1758)	+++
11. *Rutilus rutilus* (Linnaeus, 1758)	+++++
12. *Scardinius erythrophthalmus* (Linnaeus, 1758)	+++++
13. *Tinca tinca* (Linnaeus, 1758)	++
Order II. Siluriformes	
Family 3. Siluridae Cuvier, 1816	
14. *Silurus glanis* Linnaeus, 1758	+++
Order III. Gadiformes	
Family 4. Lotidae Jordan et Evermann, 1898	
15. *Lota lota* (Linnaeus, 1758)	+++
Order IV. Syngnathiformes	
Family 5. Syngnathidae Rafinesque, 1810	
16. *Syngnathus nigrolineatus* Eichwald, 1831	++++
Order V. Perciformes	
Family 6. Percidae Cuvier, 1816	
17. *Perca fluviatilis* Linnaeus, 1758	+++++
Family 7. Eleotrididae Regan, 1911	
18. *Perccottus glenii* Dubowski, 1877	+++++

Note RN—relative numbers; +—the species is represented by single individuals; ++—very low abundance (0.01–0.10%); +++—low abundance (0.11–1.00%); ++++—medium abundance (1.1–4.0%); +++++—high abundance (1.1–4.0%)

Table 18.2 Number of species belonging to different faunal complexes in the Krasnoyarsk floodplain

Name	P.-C.F	B.P	P.-C.S	T.R.F	A.F	Ch.P
Number of species	8	5	1	2	1	1

Note Faunal complexes: P.-C.F.—the Ponto-Caspian freshwater, B.P.—the Boreal plain, P.-C.S.—the Ponto-Caspian Sea, T.P.F.—the tertiary plain freshwater, A.F.—the Arctic freshwater, Ch.P.—the Chinese plain

According to Nikolsky (1951), new species are formed through adaptation to the abotic and biotic conditions of the zone of their formation, and their formation is typically of a group nature. As a result, the so-called faunal complex is formed, which is a group of species of common geographical origin. During the time of their separate existence, faunal complexes diverged in different directions and acquired not only species but also ecological specificity.

In the historical past, faunal complexes that compose the current ichthyofauna of the Volga basin were formed over large areas and in different landscape and geographical zones. Changed climatic conditions, tectonic activity, and human activity have led to and are currently leading to species dispersion. As a result, virtually all water bodies of the Volga River basin include representatives of different faunal complexes, and their ichthyofauna is based on species complexes which were formed at the location of the present water bodies, as well as those formed in the nearby territories (Shashulovsky and Ermolin 2005b).

In its historical development, the Lower Volga is closely connected to the water bodies of the Ponto-Caspian Sea. On the other hand, the middle and upper parts of the Volga River flow through the Boreal Plain Zone. Accordingly, the representatives of these complexes are the basis of the ichthyofauna of the Lower Volga water bodies.

Based on the studies by Nikolsky (1951), Nebolsina (1980) and others, the ichthyofauna of the wintering pits in the Krasnoyarsk floodplain can be divided into six faunal complexes in accordance with their morphological and ecological features (see Table 18.2).

Two of the complexes, the Ponto-Caspian freshwater and the Boreal plain, form the basis of ichthyofauna (77%).

The most diverse is the Ponto-Caspian freshwater complex and it is represented by 8 species (Table 18.2). The Ponto-Caspian fauna has an ancient history, and its components were formed at the end of the Tertiary period. According to Nikolsky (1951), representatives of the genera Rutilus, Scardinius, and Pelecus can be found in the sediments of the Pontic Sea. Modern representatives of the fauna of the Ponto-Caspian are widespread in the rivers of the Ponto-Caspian, Baltic Sea region, and Central Europe, and in the basins of the Northern Dvina and Pechora rivers. Acclimatization works increased number of representatives of this faunal complex in water bodies of Siberia.

The Boreal Plain faunal complex in the Krasnoyarsk floodplain is represented by 5 species (see Table 18.2). The Boreal Plain complex was formed in water bodies located in far-northern latitudes, which led to the development of specific features and, above all, high adaptability and resistance to adverse environmental conditions.

All of them are characterized by high euryoxicity and eurythermality. The faunal complex includes fish adapted to conditions of low oxygen content that breed at high temperatures (crucian carp), and those that breed at relatively low temperatures (pike, roach, ide). Obviously, the species of the studied complex were formed in slow-flowing water bodies with trophicity fluctuations from a high level to dystrophic one.

The species of the Boreal plain and the Ponto-Caspian freshwater complexes are widespread. Broad adaptive capabilities of the above complexes allow fish species to live in most biotopes, which ensures their dominance in ichthyocenosis.

The Tertiary freshwater complex is represented by two species: carp and catfish. This ancient group formed in the Upper Tertiary and in the past was widely distributed in water bodies throughout southern Siberia, as well as in Europe and America. Carp and catfish, similar to the previous species, can be called historically local fish. The species of the Tertiary freshwater complex are characterized by small-sized eyes and using their sense organs for food orientation. These features developed as an adaptation to high turbidity of rivers in which the complex was formed.

The Ponto-Caspian Sea complex includes the Black Sea needlefish. This species originated from marine forms but sufficiently adapted to life in fresh water, and this adaptability was developed in the geological past. At present, it is widespread in all water bodies of the Volga cascade.

The Arctic freshwater complex is represented by only one species, burbot, which is widespread in the Northern Hemisphere and leads an active lifestyle at low temperatures. In northern water bodies, this species is active throughout a year. In the water bodies of the Lower Volga, it is active in the autumn–winter–early spring period at water temperatures below 14 °C. At higher temperatures, burbot hides in the outlet of cold waters or pits with weakly heated water.

The Chinese plain complex includes one species, the Chinese sleeper. Earlier, this species did not occur in the reservoir. The first specimen was caught nearby the village of Usovka at the confluence of the Tereshka River and the Volga River in 1988, the second specimen was taken in the Krasnoyarsk floodplain in 1994, and the third one was found near the left bank opposite the city of Volsk in 1996. Since 1997, the data on the occurrence of Chinese sleeper have been reported from various parts of the reservoir (Shashulovsky 2001).

At present, the abundance of Chinese sleeper in the Krasnoyarsk floodplain is sufficiently high.

18.4 Discussion

Winter is a favorable season only for burbot among the above species since the species was formed in low temperature water bodies. It does not form winter accumulations. The behavior pattern of the species implies an active lifestyle, and its basic vital functions are realized in the late autumn, winter and early spring, when the water temperature does not exceed 10–14 °C. At higher temperature, it stops feeding and hides in shelters (in depressions of the bottom) with low water temperature.

For other fish species, winter is a difficult (unfavorable) period of the year, especially for fish from the Ponto-Caspian freshwater and the Tertiary plain freshwater complexes, since these species are formed in water bodies with a predominant water temperature of about 15–30 °C. When the water temperature drops below 10–7 °C, the fish from these faunal complexes stop feeding and form wintering aggregations. The general behavior pattern of species of the Boreal plain complex occupies, to some extent, a middle position between the behavior pattern of species from the Arctic freshwater complex on the one hand and that of the Ponto-Caspian freshwater and the Tertiary plain freshwater complexes on the other one. These species continue to be active and feed in winter, although they become less active as compared to summer. They can form winter aggregations separately or as part of wintering pits with fish species from the Ponto-Caspian freshwater and the Tertiary plain freshwater complexes.

The Volgograd Reservoir is a complex hydrological system with wintering pits in the deep-water channel part. They can be found in floodplain areas, yet their location is difficult to determine by hydrometric and bathymetric parameters due to their relatively small size. Understanding the behavior patterns of fish in winter in the context of the zoogeographic origin helps to determine the location of wintering pits ~~to~~ and preserve the biological resource potential of the water body.

18.5 Conclusion

Thus, all commercial fish in the Volgograd Reservoir, except burbot, can form wintering pits in floodplain areas. However, this behavior pattern is vital for species from the Ponto-Caspian freshwater and the Tertiary plain freshwater complexes; it is determined by the strategy developed during species formation and aimed at survival in unfavorable winter conditions.

The accumulation of fish from the Ponto-Caspian freshwater and the Tertiary plain freshwater complexes, including such valuable species as crucian carp, bream, catfish, silver bream, and asp, in the autumn–winter period confirms suitability of the study areas for their large aggregations. Therefore, this can be used as an indicator of compliance of the pits with the status of wintering pits.

References

Chemagin AA (2016) Current issues of science and education. https://www.science-education.ru/ru/article/view?Id=25254 (in Russian)

Cherepanov KM (1995) Features of the reproduction of fish stocks in the Volgograd Reservoir. Works GosNIORKh 315:47–60 (in Russian)
Ermolin VP (2010) Composition of the ichthyofauna of the Saratov Reservoir. Iss Ichthyol 50(2):280–284 (in Russian)
Nebolsina TK (1974) General characteristics of the shallow-water zone of the Volgograd Reservoir. Izv Gosniorkh 89:151–158 (in Russian)
Nebolsina TK (1980) Ecosystem of the Volgograd Reservoir and the ways of creating a rational fisheries: Dis. ... Doctor of Biol. Sciences, Saratov, RSFSR (in Russian)
Nikolsky GV (1951) On the method of research zoogeography. Iss Geogr 24:263–275 (in Russian)
Pravdin IF (1966) Guide to the study of fish. USSR, Food Industry, Moscow (in Russian)
Shashulovsky VA (2001) On the appearance of the self-climatizant Chinese sleeper in the Volgograd Reservoir. In: Shlyakhtin (ed) Fundamental and applied aspects of functioning of aquatic ecosystems (187–189). Saratov University Publishing House, Saratov, RF (in Russian)
Shashulovsky VA, Ermolin VP (2005a) Ichthyofauna composition of the Volgograd Reservoir. Iss Ichthyol 45(3):324–330 (in Russian)
Shashulovsky VA, Ermolin VP (2005b) Transformation of the structure of the ichthyocenosis of the Volga River in the ecosystem of the Volgograd Reservoir. Povolzhsky Ecol J 2:185–190 (in Russian)
Shashulovsky VA, Mosiyash SS (2010) Formation of biological resources of the Volgograd Reservoir during the succession of its ecosystem. RF, KMK, Moscow (in Russian)
Zakora LP (1974) Feeding of juvenile fish in shallow waters of the Volgograd Reservoir. Bull GosNIORKh 89:206–211 (in Russian)

Chapter 19
Ways of Penetration and Further Spread of White-Eye Bream (*Ballerus sapa*, Pallas, 1814) in Inland Water Bodies of Kaliningrad Oblast (Russia)

Oleg Novozhilov, Yulia Aldushina, and Pavel Baranovsky

Abstract In the 19the century, the white-eye bream started its spread beyond its native range, reached the basins of the White, Baltic and North Seas and in recent decades has been recorded in brackish and fresh water bodies of Kaliningrad oblast. The pathways of white-eye bream spread beyond its native range and its penetration into the water bodies of Kaliningrad oblast are analyzed in the article. The data of about seven thousand fishing operations performed using five different fishing gear have been analyzed. From 2009 to 2017, solitary specimens of white-eye bream were annually recorded in the Pregolya River of the Vislula Lagoon basin along a 10–80 km distance from its mouth, and in 2017, it was captured in the Neman River of the Curonian Lagoon basin, 80 km from its mouth. White-eye bream in water bodies of Kaliningrad oblast has an average length of 21.5 cm and average weight of 184 g in the Pregolya River and a single specimen captured in the Neman River had a length of 18.5 cm and a weight of 111 g. White-eye bream penetrated to the Baltic Sea basin via the central European invasion corridor at the end of the twentieth century and since 2009, it has reached large rivers of Kaliningrad oblast probably via the Vistula Lagoon, namely the Vistula River basin. Further spread of white-eye bream is possible upstream of the Pregolya River basin but can be limited by a hydraulic structure and along its tributary (the Lava River). The distribution in the Neman River basin can occur in two directions, upstream and along its major tributaries flowing through the territories of the Republic of Lithuania and the Republic of Belarus.

Keywords White-sea bream · Spread invasion corridors · Pergolya River basin · Neman River basin

O. Novozhilov (✉) · Y. Aldushina · P. Baranovsky
Kaliningrad State Technical University, Sovetsky Avenue, 1, 236022 Kaliningrad, Russia
e-mail: oleg.novozhilov@klgtu.ru

A. G. Arkhipov (ed.), *Sustainable Fisheries and Aquaculture: Challenges and Prospects for the Blue Bioeconomy*, Environmental Science and Engineering, https://doi.org/10.1007/978-3-031-08284-9_19

19.1 Introduction

The white-eye bream (*Ballerus sapa* (Pallas, 1814)) is a closely related species of bream and a rather large fish of a length up to 40 cm (Reshetnikov 2002; Spotlight Resources n.d.), associated with deep river sections in its native range and, despite the formation of semi-anadromous forms in the lower reaches of large rivers, unlike bream, it prefers fresh waters. It avoids water bodies with a slow flow and is absent in floodplain water bodies, lakes, small rivers (Reshetnikov 2002; Zhukov 1965, 1988). It should also be mentioned that before the nineteenth century, the range of white-eye bream was limited to the Ponto-Caspian basin, which is its natural range (Kazarinov et al. 2021).

Since the nineteenth century, white-eye bream has started its spread in the northern direction, to the basins of the White, Baltic and North Seas (Zhukov 1965; Novoselov 2000; Alimov and Bogutskaya 2004; Terlecki 1990; Kakarenko et al. 2008; Skóra and Skóra 2013; Bischoff et al. 1998). In the Baltic Sea basin, it has reached brackish-water coastal water bodies, namely, the practically freshwater Curonian Lagoon and the brackish-water Vistula Lagoon, which makes it possible to predict its appearance in the freshwater water bodies of Kaliningrad oblast, which are directly connected to the abovementioned lagoons.

Despite the high-intensity monitoring studies in the water bodies of Kaliningrad oblast conducted by the Kaliningrad State Technical University for more than 20 years, the white-eye bream had not been recorded in catches until 2009; later it has become a rare but rather stable component of the ichthyofauna in a number of water bodies of Kaliningrad oblast.

The aim of this work is to analyze the routes of white-eye bream spread beyond its native range, to determine the ways of its penetration into the water bodies of Kaliningrad oblast and to predict its possible further spread.

19.2 Materials and Methods

The work is based on the data of monitoring studies of the Department of Ichthyology and Ecology, Kaliningrad State Technical University. The regular observations in the Vistula Lagoon basin in Kaliningrad oblast have been conducted since 1996, and since 2007, they have also started in the basin of the Curonian Lagoon. The number of fishing operations was about 200 per year in the period from 1996 to 2009.

During 2009 to 2017, the studies in the inland waters of Kaliningrad oblast were intensified; about 4.5 thousand catches were performed using five types of fishing gear, on average about 500 catches per year (see Table 19.1).

Control catches were conducted using the following fishing gear:

- fry dragnet of 10 m long with a 4 mm mesh codend;
- beach seine of 120 m long with a 12 mm mesh codend and tow line length of 150–300 m;

Table 19.1 Number of catches in water bodies of Kaliningrad oblast

Water body	Years									Total
	2009	2010	2011	2012	2013	2014	2015	2016	2017	
Vistula lagoon basin										
Vistula lagoon	19	158	–	102	80	8	55	–	78	500
Pregolya river	139	1	–	121	–	201	97	185	146	890
Curonian lagoon basin										
Curonian lagoon	117	138	76	3	–	–	–	–	–	334
Neman river	120	55	208	406	198	411	319	510	332	2559
Deyma river	105	–	–	80	–	–	–	–	–	185
Total	500	352	284	712	278	620	471	695	556	4468

Source Our data

- drift nets of 30–60 m long with a mesh size of 14–70 mm;
- set gillnets of 15–60 m long with a mesh size of 12–70 mm;
- handy electric fishing device.

The catches were carried out in the estuary area of the Pregolya River in the Vistula Lagoon in the summer (June–July). The studies in the lower reaches of the Neman River (at stations up to 80 km from the mouth) were conducted from April to October. Fish were caught both in the riverbed and in floodplain water bodies (backwaters). Studies in the Pregolya River were carried out in its lower course, along the river from the city of Kaliningrad to the city of Gvardeysk (10–80 km from the mouth) and in floodplain water bodies (Lake Pustoe, Lake Voronye, Sokolniki quarry) from April to October. The studies in the Deyma River were carried out in its upper (city of Gvardeysk) and lower (city of Polessk) reaches.

Fish samples were collected and processed according to the common methods (Pravdin 1966). The length was measured with accuracy to within one millimeter, the weight, to a gram. The commercial length (to the end of the scale cover) and the total weight were measured.

19.3 Results

White-eye bream was first caught in the Pregolya River basin in Kaliningrad oblast in 2009, and in 2017, in the Neman River basin (see Table 19.2). The size and weight characteristics of white-eye bream varied between 18 and 26 cm and between 111 and 372 g, respectively.

After the first capture of white-eye bream in the Pregolya River basin in 2009, now it is consistently found in catches almost annually at a distance of 10–40 km from the mouth of the Pregolya River. In the Neman River, it was recorded 8 km from the mouth of the river.

Table 19.2 Length–weight parameters of white-eye bream from water bodies of Kaliningrad oblast

Water body	Parameters	Years						Total
		2009	2012	2014	2015	2016	2017	
Pregolya river	N, items	1	1	5	1	2	–	10
	L_{avg}, cm	26.5	19.5	20.0	24.5	22.3	–	21.5
	W_{avg}, g	372	107	149	244	188	–	184
Neman river	N, items	–	–	–	–	–	1	1
	L_{avg}, cm	–	–	–	–	–	18.5	18.5
	W_{avg}, g	–	–	–	–	–	111	111

Source Our data

19.4 Discussion

A number of Ponto-Caspian species enter the Baltic Sea basin via invasion corridors, among which central, northern and southern ones are distinguished (Bij de Vaate et al. 2002). The expansion of the range of such fish species as round goby occurred via these pathways (Tylik 2003; Rizevsky et al. 2016).

The spread of white-eye bream along the northern invasion corridor has started earlier, since the end of the nineteenth century. The first finding of white-eye bream beyond the Black Sea-Caspian Sea and Sea of Azov basins was reported in the publications of Varpakhovsky (1898) and Sabaneev (1911). In the authors' opinion, white-eye bream penetrated to the Volkhov River from the upper reaches of the Volga River. The next finding of white-eye bream beyond its native range was recorded in the Vychegda River, a tributary of the Severnaya Dvina River in the basin of the White Sea, in 1975 (Novoselov and Studionov 2000).

In the twentieth century, the construction of hydraulic facilities, climate change, intensive shipping led to a change in natural biodiversity and transformation of natural ecosystems. Global warming is considered a factor of successful dispersal of white-eye bream in the Northern Dvina basin (Alimov and Bogutskaya 2004).

Despite the milder climate in the Baltic Sea basin compared to the White Sea and a hydraulic connection with the Ponto-Caspian Basin (the Dnieper-Bug Canal 1783), the dispersal of white-eye bream in the Baltic Sea basin started via the central invasive corridor only at the end of the twentieth century.

The spread of white-eye bream in this direction began in 1989, when it was found in the reservoir on the Bug River (Terlecki 1990). Around the same time (1985), white-eye bream was found in the Vistula River (Kakarenko et al. 2008). The spread rate of white-eye bream in the Bug-Vistula river system was relatively low and amounted to 400 km for 20 years (Skóra and Skóra 2013). At the beginning of the 21th century (2006–2007), white-eye reached the Baltic Sea basin, namely the Gulf of Gdansk (Skóra and Skóra 2013) (see Fig. 19.1).

Along the southern invasion corridor and the Rhine River basin, white-eye bream began to penetrate into the North Sea basin in 1995 at a rate of 190 km per year (in

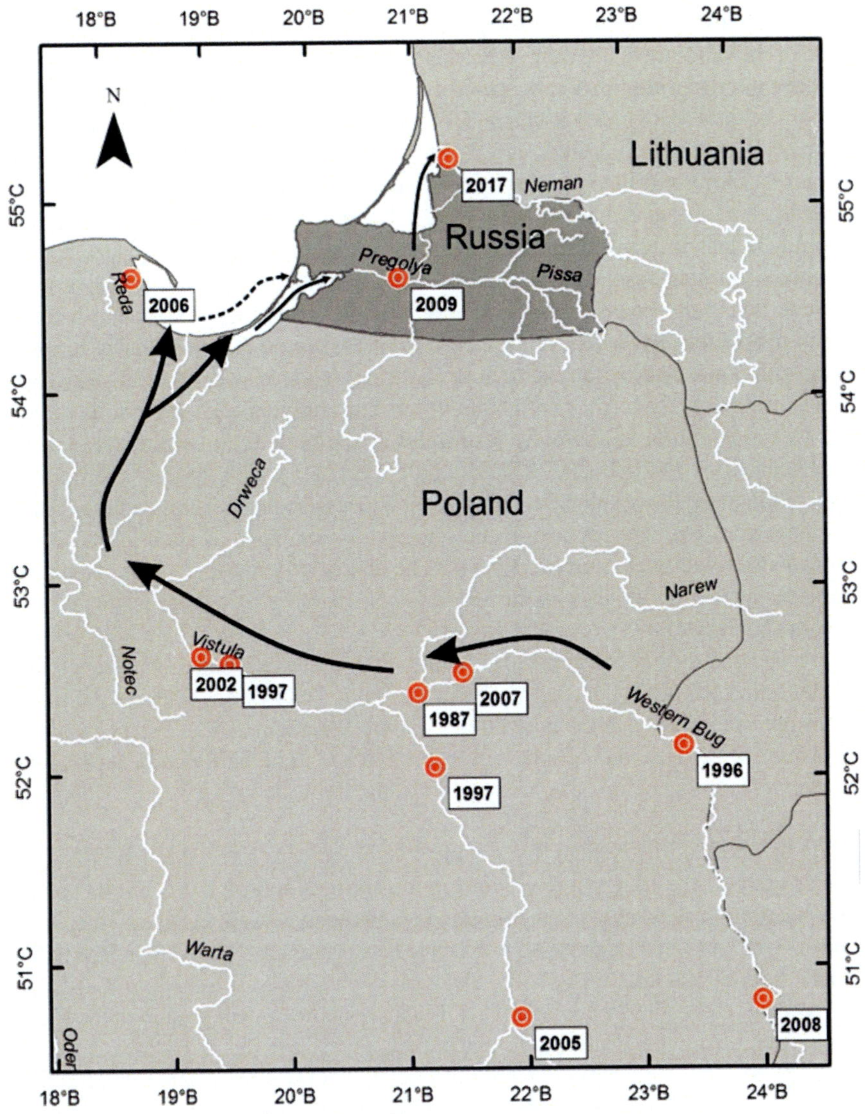

Fig. 19.1 Scheme of white-eye bream penetration into the Baltic sea basin (*Source* Catches in Poland presented by Skóra and Skóra (2013), in Russia—our data)

the first year), and 100 km in the second year (Bischoff et al. 1998). Thus, by 2017, white-eye bream had been found in the basins of the White, Baltic and North Seas.

Since 2009, white-eye bream has been recorded within the territory of Kaliningrad oblast and 11 specimens of white-eye bream were caught during the period of monitoring studies.

Most of the specimens were caught in the area of the former quarry for the extraction of the sand-gravel mixture, connected to the Pregolya River by several channels in the autumn period. A solitary specimen was caught at a distance of 15 km from the mouth in the Pregolya River channel. Despite regular fishing in the Vistula Lagoon (the near-mouth area of the Pregolya River), white-eye bream was not found there which is also confirmed by surveys of fishermen engaged in industrial fishing. It should be mentiones, that morphological differences between white-eye bream and bream are insignificant for fishermen and they did not identified their species.

In our opinion, despite the absence of white-eye bream in catches in the lagoon, its penetration to the Pregolya River occurred via the Vistula Lagoon, where it entered from the Vistula River through the system of the Nogat and Szkarpawa branches. This also confirmed by the fact that the hydrographic network connects the Vistula River basin with the territory of Kaliningrad oblast only through the Vistula Lagoon.

This assumption is indirectly confirmed by the presence of white-eye bream in the Reda River in 2006–2007 (Skóra and Skóra 2013). The distance from the site where the Reda River splits from the Nogat River is about 100 km, and the distance to the site of the first capture in the Pregolya River basin is about 180–200 km, which she could overcome for 3 years. The ability of white-eye bream to perform long-distance migrations is confirmed by other literature data (Bischoff et al. 1998; Skóra and Skóra 2013).

At the same time, it cannot be exclude that white-eye bream can penetrate to Kaliningrad oblast via the coastal area of the Baltic Sea as well, where water salinity does not exceed $7^{0}/_{00}$ in the horizon of 0–10 m (Dubravin 2017). The possibility of this route of penetration is confirmed by the spread of another Ponto-Caspian representative, round goby in the coastal waters of the Baltic Sea (Tylik 2003; Kodukhova et al. 2017; our unpublished data).

In the basin of the Neman River, white-eye bream was recorded in the floodplain water body of the Skirvyte River during the spring flood of 2017. In our opinion, the penetration occurred via the following pathway: the Pregolya River—the Deyma River—the Curonian Lagoon—the Skirvyte River (branch of the Neman River). This route is about 100 km.

Another theoretically possible way of white-eye bream penetration to the Neman River basin from the Vistula River basin may be through the Augustow Canal with a length of more than 100 km (Noviczkij et al. 1911). However, regular studies in the Neman River upstream of the delta system have shown the absence of white-eye bream in the river both in the territory of Russia and Lithuania.

Thus, the conclusion can be made that white-eye bream has currently penetrated from the Vistula River basin to the basins of the Pregolya and Neman rivers, which is a transboundary body of water connecting Russia, the Republic of Lithuania and the Republic of Belarus. The biological features of the species make it possible to predict its distribution only in the largest rivers of Kaliningrad oblast, the Pregolya River and the Deyma River. In the Pregolya River, white-eye bream can reach the city of Chernyakhovsk, located at a distance of about 100 km from the mouth or 70 km from the site of capture, where its distribution will be limited by a dam. In the Lava River, the tributary of the Pregolya River, through which white-eye bream could theoretically penetrated to the system of the Masurian lakes, there are also dams that almost completely exclude this pathway of the species spread.

In our opinion, further spread of white-eye bream in the Neman River basin, will occur along the main channel to the city of Kaunas (Lithuania), where the river is blocked by the dam of the Kaunas hydroelectric power plant and into the tributaries, the Sheshupe River in the territory of Kaliningrad oblast, the Jura (the length of 177 km), Nyavezis (the length of 210 km), Viliya (the length of 500 km), Minija (the length of 200 km) rivers on the territory of the Republic of Lithuania and further to the Republic of Belarus. The penetration along the Baltic Sea coast in the northern direction to more desalinated waters also remains possible.

The available literature data show that the introduction of white-eye bream into new water bodies can cause tense relationships with native species, in particular in the Northern Dvina River; the index of food similarity of white-eye bream and whitefish is 62.8%, and the index of overlapping of food niches is 89.9% (Novoselov 2020). Thus, in view of the dominance of benthic fish in the commercial ichthyofauna (Hlopnikov 1992), the appearance of a new food competitor can lead to changes in the sustainability of industrial fisheries.

19.5 Conclusion

The transformation of aquatic ecosystems in the Baltic Sea basin occurs under the effect of the penetration of a number of Ponto-Caspian species. Following the round goby, another benthivorous species, white–eye bream, has naturalized in the area of the southern Baltic Sea, which may lead to a change in the stability of the fishery. Its biological features make it possible to suggest its further spread along the river systems of the Pregolya River and Neman River basins, as well as its subsequent penetration northwards along the sea coast. The further expansion of the natural range, in addition to the ongoing processes of global warming, can be facilitated by the potential implementation of the E 40 Inland Waterway, a transnational project of the European Union aimed at the connection of the Vistula water systems with the Dnieper-Bug canal [United Nations Economic Commission for Europe (UNECE), ECE/TRANS/SC.3/WP.3/2016/13, 2016].

References

Alimov AF, Bogutskaia NG (eds) (2004) Biological invasion in water and terrestrial ecosystems. Tovarishestvo nauchnyh izdaniy KMK, Moskow (in Russian)

Bij de Vaate A, Jażdżewski K, Ketelaars HAM, Gollasch S, Van der Velde G (2002) Geographical patterns in range extension of Ponto-Caspian macroinvertebrate species in Europe. Can J Fish Aquat Sci 59:1159–1174

Bischoff A, Freyhof J, Staas S (1998) Records of the White-eye bream Abramis sapa (Pallas 1811) (Teleostei: Cyprinidae) in Rhien. Lauterbornia 33:5–8

Dubravin VF (2017) Evolutions of thermohaline structure fields in the Baltic Sea waters. Publisher House "Pero", Moskow (in Russian)

Hlopnikov MM (1992) Food relations of benthos-feeding fishes of Vistula lagoon. In: Hlopnikov MMM (ed) Ecologycal and fishery investigations in Vistula lagoon of Baltic Sea. Atlantic Science Investigation Institute of Fish industry and Oceanography, Kaliningrad, pp 165–188 (in Russian)

Kakareko T, Kobak J, Terlecki J, Hadowski M (2008) External morphology and growth rate of white-eye bream Ballerus sapa (Cyprinidae, Teleostei) in a lowland dam reservoir on the lower Vistula River (Włocławek Reservoir, central Poland). Folia Zool 57:435–451 (in Russian)

Kazarinov SN, Merzlyakov IN, Ponosov SV, Komarova LV (2021) Species composition and distribution features of the ichthyofauna of the Kama reservoir. Vestnik Permskogo universiteta. Biologija 1:39–52 (in Russian)

Kodukhova YuV, Borovikova EA, Ezhova EE, Gushchin AV (2017) Features of the morphology of round goby (Neogobius melanostomus) in the South-Eastern Baltic Sea. Reg Ecol 3:65–74 (in Russian)

Novitsky VF (ed) (1911) Military encyclopedia, vols 1–18. Tovarishestvo I. D. Sytina, Sankt-Petersburg (in Russian)

Novoselov AP, Studionov II (2000) The composition of ichthyofauna of estuary zone of Northern Dvina River. Iss Fish 1(2–3):75–76 (in Russian)

Novoselov AP (2020) New species of fish in reservoir of Europian North-East of Russia. Ecology 6:457–464 (in Russian)

Pravdin I (1966) Guide to the study of fish (mainly freshwater). Food industry, Moscow (in Russian)

Reshetnikov YuS (ed) (2002) Atlas of Russian freshwater fishes, vols 1–2. Nauka, Moskow (in Russian)

Rizevsky VK, Ermolaeva IA, Leschenko AV, Hryhorchyk AP (2016) Ponto-Caspian autoinvaders in the structure of juvenile fishes in coastal shallow water area within Belorussian section of central invasion corridor. Iss Fish Ind Belarus 32:206–219 (in Russian)

Sabaneev LP (1911) Fish of Russia: life and angling of local freshwater fish. A. A. Kartsev, Moscow (in Russian)

Skóra M, Skóra K (2013) First report on the white-eyed bream Ballerus sapa (Pallas, 1811) in the Reda River, a tributary to the Baltic Sea. J Appl Ichthyol 29:930–932 (in Russian)

Terlecki J (1990) White-eye bream, Abramis sapa (Pallas, 1811)—first record in Poland, in man-made Zegrzyńskie Lake. Acta Ichthyol Piscat 22(2):97–108 (in Russian)

Tylik KV (2003) Roun goby distribution in the Baltic Sea coastal zone of Kaliningrad oblast. In: Matishov GG (ed) Evolution of marine ecosystems under influence of invaders and artificial mortality of fauna: International conference, Azov 15–18 June. Murmansk Marine Biological Institute of Kola Scientific Center of Russia Academic Science, Rostov-on-Don (in Russian)

United Nations Economic Commission for Europe (UNECE). Inland Transport Committee (2016) Forty-ninth session Geneva, 22–24 June 2016 Item 5 (b) of the provisional agenda. Inland waterways infrastructure: Inventory of Main Standards and Parameters of the E Waterway Network ("Blue Book"). Development of the E 40 international waterway on the Dnieper-Vistula Section (ECE/TRANS/SC.3/WP.3/2016/13). https://unece.org/fileadmin/DAM/trans/doc/2016/sc3wp3/ECE-TRANS-SC3-WP3-2016-13r.pdf (in Russian)

Varpahovskij NA (1898) Qualifier of freshwater fishes of European part of Russia. A. A. Suvorin, Sanct-Petersburg (in Russian)

White-eye bream—Abarmis sapa (Pallas. 1814) (n. d.) Waarneming. ttps://waarneming.nl/species/22437/observations/?date_after=2000-01-01&date_before=2021-11-02&province=&search=&user=&location=&sex=&life_stage=&activity=&method. Accessed 2 Dec 2021 (in Russian)

Zhukov PI (1965) Fishes of Belorussia. Nauka I Tehnica, Minsk, USSR (in Russian)

Zhukov PI (1988) Handbook of freshwater fishes ecology. Nauka I Tehnica, Minsk, USSR (in Russian)

Chapter 20
Using the Gill Nets Survey for Assessment of Fish Stock and Allowable Catch in the Vistytis Lake, Kaliningrad Oblast, Russia

Sergey Shibaev, Oleg Novozhilov, and Pavel Baranovsky

Abstract The paper describes a method to estimate the abundance of benthic fishes in the Vistytis Lake (area of 1.8 thousand ha, maximum depth of 52 m) using bottom gill nets with 12–60 mm mesh size. Annual 10-day surveys were performed from 2007 to 2020 at 30 stations located throughout the water area. The gill nets are not suitable for assessment of absolute number of fishes since it is unknown catching square of the net and catchability coefficient. However, the integration of catches per unit effort (in number of fishes or in kilograms per 25 m of net length per fishing day) and the area of the respective depth ranges and mesh size, can be used to calculate the relative abundance and biomass of the stock as well as the species composition of ichthyocenosis. The net mesh size can be considered as an identifier of the fish size group. This gives an opportunity to analyse the length structure of the population. The catches per unit effort for each mesh size can be interpreted as a total allowable catch (TAC), provided each hectare of the water body is fished once a year with a 25 m long net with each mesh size. The average stock abundance estimated with application of this approach gives the number of fishes 681.9 ± 144.8 thousand specimens and 38.44 ± 7.61 tons correspondingly, and the interannual variation coefficient for different species ranges from 50 to 100%. With regard to the water body area, the fish population density is 378 specimens/ha and 21.7 kg/ha, which is comparable to the ichthyofauna characteristics of the oligotrophic lakes in northwestern Russia. The method can be applied to any water body inappropriate for fishing by gears such as a trawl to assess the numbers of fish.

Keywords Fish stock assessment · Gill nets · Mesh size · Demersal ichthyocenosis · Species composition · Small scale fisheries · Allowable catch · The Vistytis Lake · Spatial distribution · Population length structure

---

S. Shibaev (✉) · O. Novozhilov · P. Baranovsky
Kaliningrad State Technical University, Sovetsky avenue, 1, 236022 Kaliningrad, Russia
e-mail: shibaev@klgtu.ru

A. G. Arkhipov (ed.), *Sustainable Fisheries and Aquaculture: Challenges and Prospects for the Blue Bioeconomy*, Environmental Science and Engineering, https://doi.org/10.1007/978-3-031-08284-9_20

20.1 Introduction

The estimation of fish stocks in exploited small lakes, which is mandatory for the calculation of the total allowable catch, is specific and differs significantly from the approaches used for large inland waters. These peculiarities related to the impossibility of using active fishing gear such as a trawl or seine with known catching square to estimate fish population density, high variability of biotopes and therefore fluctuation of the spatial distribution of fish resulting in a large statistical error in mean values. In addition, each fish species may have different patterns of behavior, catchability and selectivity towards the fishing gear used, which also contributes to uncertainty in stock assessment.

A typical example of a small exploited water body is the Vistytis Lake, which is important for local small-scale fisheries in the Kaliningrad region, Russia. Preliminary studies have shown that active fishing methods cannot be used to assess fish abundance due to the specific morphology of the water body, spatial–temporal distribution of fish, and high water transparency, which allows fish to see and actively avoid moving fishing gear such as trawls or seines.

The only fishing gear that allows effective catch of any commercial fish species are the gill nets, which can be used anywhere in the water. Nevertheless, nets do not allow the catching square to be taken into account and therefore do not provide an opportunity to estimate the absolute abundance of the stock (Shibaev 2014).

The objective of this study is to develop a methodology for estimation of the status of stocks and allowable catch according to the relative indicators obtained using gill nets catches.

20.2 Materials and Methods

The investigation of the Vistytis Lake from 2007 to 2020, which was aimed at assessing the fish stock by means of the bottom gill nets, provided material for this study. A set of gill nets with mesh size 12, 14, 16, 18, 20, 22, 24, 30, 35, 40, 45, 50, 55, and 60 mm was used. The net length varied from 10 m for small-mesh nets up to 50 m for large-mesh nets, and was selected to make the catch sufficient to estimate the species and length composition of catch and to avoid excessively big catch which takes much processing time. The net height was 1.8–2.3 m also depending on mesh size.

Scientific catching was carried out at 30 standard stations located in a different depth. The depth interval was 5 m and the number of catchings for each depth range was up to 5 depending on the range area. Thus on each station up to 15 catchings were performed for each mesh size (12–60 mm). Fishing duration was about 24 h to avoid the impact of daily fish migrations. The survey lasted up to 10 days annually and the total number of catchings reached 450. The result of survey was a mass of catches per unit effort (CPUE) for each gill net. These data were used for estimation of fish stock and allowable catch of each fish species (Shibaev, Sokolov, 2014).

The total number of scientific catchings performed to estimate the stock amounted to 6.8 thousand, the number of fish involved for mass measurements was 114 thousand. The data were processed using the Rybvod information and analytical system that allowed automatic analysis of a considerably large bulk of information on scientific catching (Shibaev 2004).

20.3 Results

The Vistytis Lake is an oligotrophic water body with a total area of 1.8 thousand hectares. It is about 8 km long from north to south and 4 km from west to east. The lake is transboundary: the western part is located within the Russian Federation (81% of the area), and the eastern part belongs to the Republic of Lithuania (19% of the area) (Vistytis Lake 2008). The presence of three deep-water basins, narrow shallow-water zones in most parts of the water body and small macrophyte areas determine the spatial distribution of fish. The shallow water zone with depths of up to 10 m is about 30% of the lake area, while the zone with depths of 10–40 m covers about 60% of the area (see Fig. 20.1). Roach and perch are confined to the shallow water area and migrate daily to the shore at night and offshore during the day. The

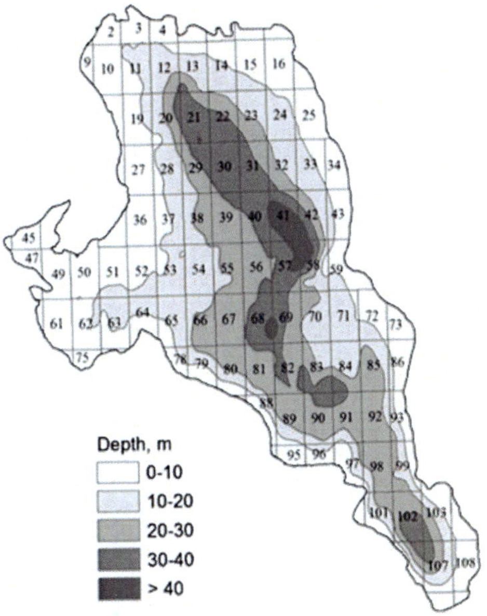

Depth, m	Square	
	ha	%
0-5	433,2	24,1
5-10	258.1	14.3
10-15	257.8	14.3
15-20	218.1	12.1
20-25	218.7	12.2
25-30	182.6	10.1
30-35	108.1	6.0
35-40	96.5	5.4
40-45	22.8	1.2
45-50	4.9	0.2
50-55	0.1	0.1
Total	1800,9	100,0

Fig. 20.1 Morphological properties of Lake Vistytis (compiled by the authors)

deeper part is inhabited by whitefish and burbot, which prefer colder water below the thermocline. Pike and tench are associated with the macrophyte beds in the bays or in the coastal zone at shallow depths, although they can occasionally occur in the open part of the lake. The maximum depth of the water is 52 m, resulting in the formation of two relatively isolated fish communities—the pelagic community which for over 90% consists of vendace, dweller of the water layer from 15 to 45 m, and the demersal one (Vistytis Lake 2008; Shibaev, Aldushin 2011; Shibaev et al. 2017). The demersal ichthyocenosis includes seven species: roach, perch, whitefish, tench, pike, burbot and bream, which are caught with bottom nets. However, typical pelagic species such as vendace and bleak, the latter particularly at shallow depths, can also be met in bottom catches (see Table 20.1). The distribution of fish in the lake is extremely irregular and is subject not only to seasonal but also to significant daily fluctuations due to weather conditions (Shibaev et al. 2008). This makes it

Table 20.1 Species composition of gill net catches (mesh size 12–60 mm) in the Vistytis Lake (2007–2020), frequency of occurrence, % (authors' data)

Species	Mesh size, mm								
	10–15	15–20	20–25	25–30	30–35	35–40	40–45	45–50	50–60
Perch (*Perca fluviatilis* L.)	52.3	52.6	59.0	59.7	60.7	33.1	32.9	14.3	9.3
Ruff (*Gymnocephalus cernuus* L.)	76.1	71.0	42.2	16.5	12.4	5.2	2.6	2.7	1.3
Roach (*Rutilus rutilus* L.)	19.8	23.4	28.7	33.5	35.9	27.1	18.4	13.9	2.7
Whitefish (*Coregonus lavaretus* L.)	0.8	2.7	9.5	11.0	19.6	21.8	42.1	30.4	25.3
Vendace (*Coregonus albula* L.)	51.6	28.6	9.2	5.8	5.6	2.1	3.9	4.2	1.3
Pike (*Esox lucius* L.)	2.1	3.2	9.0	13.3	19.7	16.0	7.9	17.8	5.3
Burbot (*Lota lota* L.)	6.7	7.0	8.5	10.3	11.6	6.7	10.5	5.7	9.1
Bream (*Abramis brama* L.)	1.0	1.6	2.4	3.1	4.4	4.4	6.6	6.1	4.5
Tench (*Tinca tinca* L.)	1.8	0.5	0.5	0.2	0.9	2.4	3.9	5.7	0.0
Bleak (*Alburnus alburnus* L.)	8.6	2.4	0.2	0.0	0.1	0.0	0.0	0.0	0.0

very difficult to estimate the abundance of fish, which is dependent on the survey conditions each particular year.

The main commercial fishing gear currently used in the lake are gill nets with a mesh size of 16–18 (pelagic nets for vendace), 40–50 and 70 mm (bottom nets), although in the 1970s–1980s an important gear was 500 m seine, which was used for roach, perch and whitefish. For spatial management of fisheries, the water is divided into quadrats. Fishing takes place in the northern and central parts. In the bay (locally called the Tikhaya Bay) located on the western coast, the main spawning grounds of phytophilous fish, commercial fishing is prohibited. The historical maximum yield since 1962 was about 30 tons. Fishery regulation based on Total Allowable Catch (TAC), which should be estimated annually for each species with regard to the current state of its stock.

The stock status is assessed based on annual surveys performed in June–July with bottom gill nets. The calculation algorithm is as follows. Each catch per effort can be presented as amount of fish YN (specimens) or weight YW (kg)

$$Y_N[i, d, f, L, h] \tag{20.1}$$

$$Y_W[i, d, f, L, h] \tag{20.2}$$

with the following identifiers:

i—a catch number;

d—net mesh size in millimeters;

f—fish species;

L—fish length group in centimeters;

h—station depth.

Thus, summarizing the data on all catches during the survey enables assessment of fish abundance in the water body in terms of numbers or biomass. The following should be considered.

1. Each of the fish species has its own parameters of selectivity and catchability relative to the mesh size of a net. The general dependence of fish length on the mesh size is described by the linear function L = bd, where b is the species-specific coefficient; however, this dependence can be violated due to morphological characteristics of different species. For example, it is obvious that owing to the presence of spines and spiny rays of fins the catchability and selectivity ogive of perch should be different from that of spine-free species like roach.

On the other hand, the size of the fish caught is directly related to the mesh size, but the range of length fluctuation in each net is only a few centimeters. In this regard, mesh size seems to be an appropriate identifier of fish size to analyse the population structure. In this case, generalizing the catch data of the entire range of meshes provides information on some certain virtual size population structure.

2. The second feature is the need to consider the spatial distribution that is specific to each species. For example, tench is usually found in shallow water, while burbot

occurs at depths over 30 m. This makes it expedient to calculate the statistical abundance of the fish population separately for each mesh size and the depth range. Subsequently the average fish catch by the net with mesh size (d) in a given depth range (h) is the abundance index BN or the biomass index BW of a given species (f):

$$B_N[d,\ f,\ h] = \frac{\sum_{i=1}^{N_h} Y_N[i,\ d,\ f,\ h]}{N_h} S_h \tag{20.3}$$

$$B_W[d,\ f,\ h] = \frac{\sum_{i=1}^{N_h} Y_W[i,\ d,\ f,\ h]}{N_h} S_h \tag{20.4}$$

where: N_h is the number of stations in the given depth range; S_h is the area of the given depth range.

The basic statistics like standard deviation, mean error and confidence interval can be calculated in a similar way.

The total stock of fish species (f) will consist of number and biomass calculated for each mesh size (d) and for each depth range (h):

$$B_N[f] = \sum_{h_{\min}}^{h_{\max}} \sum_{d_{\min}}^{d_{\max}} B_N[d,\ f,\ h] \tag{20.5}$$

$$B_W[f] = \sum_{h_{\min}}^{h_{\max}} \sum_{d_{\min}}^{d_{\max}} B_W[d,\ f,\ h] \tag{20.6}$$

As an example, the Table 20.2 shows the calculations of the perch biomass taking into account the population length structure, which is represented as catches of nets with different mesh sizes with 5 cm interval, and the spatial distribution of 5 m over

Table 20.2 An example of calculation of perch biomass in the Vistytis Lake (2009), kg (authors' data)

Depth, m	Mesh size, mm							Total
	10–14	15–19	20–24	25–29	30–34	40–44	45–60	
0–5	117.9	180.2	582.8	471.1	964.1	194.7	8.2	2518.9
5–10	133.5	387.3	627.8	728.4	588.1	53.1	22.8	2541.1
10–15	277.4	2678.6	689.4	1748.0	1175.2	249.3	6.6	6824.5
15–20	188.3	70.4	132.3	366.0	185.3	29.5	28.4	1000.3
20–25	25.7	18.3	155.0	255.7	225.3	40.7		720.7
25–30			8.5		32.0			40.5
30–35	0.9	4.7	10.6	27.5	38.1			81.8
35–40			2.7					2.7
Total	743.8	3339.5	2209.1	3596.7	3208.1	567.3	66.0	13,730.5

the depth ranges. The data on survey catches in 2009 were considered. According to the data, the size groups of fish caught by nets with a mesh size of 25–40 mm, which were found at depths of 10–15 m, provided the largest stock biomass (1.75 tons). With regard to the abundance of all age groups, the total biomass in this zone amounted to 6.8 tons. In deeper waters, the perch biomass decreases, and at a depth exceeding 45 m, it declines to only 66 kg. Summing the catches over all depths and all applied mesh sizes gives a total stock biomass of 13.73 tons.

20.4 Discussion

Let us consider the results of assessing the abundance of fish in the Vishtytis Lake using the above algorithm. It was found that over the 14 year period of investigation (see Fig. 20.2) the estimated biomass of the stock varied over a very wide range from 70 tons in 2008 to 23 tons in 2010. At the same time, the interannual variations in the species structure of ichthyocenosis were not so significant (see Fig. 20.2). All this may indicate that the observed fluctuations in the density of the fish population are associated with the survey conditions in each particular year, since no significant changes in the habitat of fish in the lake were observed, and the fishing intensity here has been quite low over the last 20 years. In this regard, it seems necessary to use the average long-term estimates of the stock in order to characterize the ichthycenosis of the lake as a whole.

The calculation of ichthyofauna parameters has shown that the average stock abundance and biomass amounts to 681.9 ± 144.8 thousand specimens and 38.44 ± 7.61 tons correspondingly (see Table 20.3), and the interannual variation coefficient

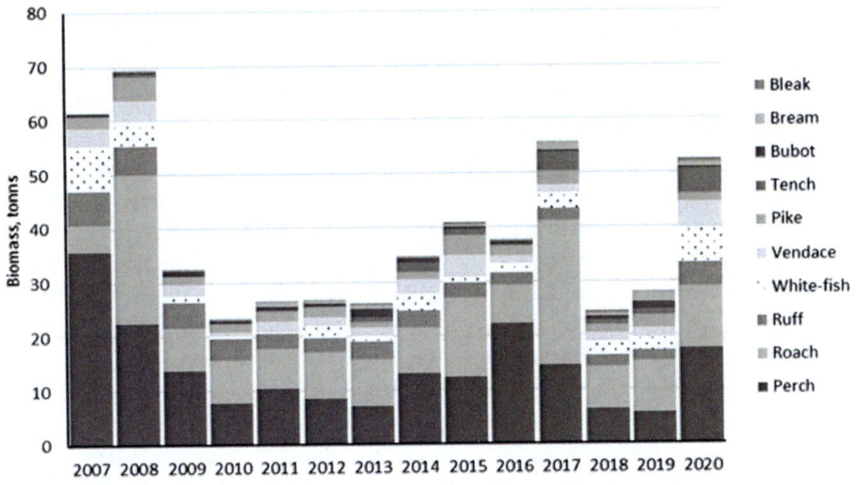

Fig. 20.2 Dynamics of demersal ichthyocenosis biomass

Table 20.3 Average annual statistical characteristics of the fish stock in the Vistytis Lake (2007–2020) (authors' data)

Species	Numbers, thousand					Biomass, ton				
	x	σ	v	Δ	%	x	σ	v	Δ	%
Perch	171.7	101.3	0.59	53.0	25.2	13.99	7.96	0.6	4.17	36.4
Roach	143.6	144.4	1.01	75.6	21.1	11.27	6.65	0.6	3.48	29.3
Ruff	248.6	84.2	0.34	44.1	36.5	3.45	1.29	0.4	0.68	9.0
Whitefish	5.8	5.0	0.86	2.6	0.9	2.76	2.17	0.8	1.14	7.2
Vendace	94.4	49.5	0.52	25.9	13.8	2.36	1.22	0.5	0.64	6.1
Pike	2.9	1.4	0.48	0.7	0.4	2.00	0.90	0.5	0.47	5.2
Tench	1.0	0.9	0.92	0.5	0.2	1.06	1.34	1.3	0.70	2.8
Burbot	2.1	0.9	0.44	0.5	0.3	0.74	0.36	0.5	0.19	1.9
Bream	2.0	1.4	0.72	0.8	0.3	0.58	0.48	0.8	0.26	1.5
Bleak	9.8	11.3	1.15	5.9	1.4	0.22	0.23	1.0	0.12	0.6
Total	681.9	276.5		144.8	100.0	38.44	14.53		7.61	100.0

x—average, σ—standard deviation, v—coefficient of variation, Δ—confidence interval

for different species ranges from 50 to 100%. With regard to the water body area, the fish population density is 378 specimens/ha and 21.7 kg/ha, which is comparable to the ichthyofauna characteristics of the oligotrophic lakes in northwestern Russia (Kitaev 2007).

Pursuant to the multi-year averages, the main fish species that form the abundance of ichthyofauna are perch, roach and ruff, while roach and perch dominate in biomass and provide more than half of the stock. Commercially valuable species such as whitefish, pike, burbot, and bream make up about 20% of biomass. At the same time, as exactly these species are the most caught ones, the quota in recent years was used up to 100% while quotas for perch and roach remained significantly underutilized.

The results not only provide an estimate of the stock status, but also an assessment of possible fish yield in the water body. This may be based on the following considerations. Abundance and biomass showed in the Table 20.3 are calculated as the catch per unit effort, per 25 m of the net per fishing day multiplied by the water body area. This value can be interpreted as the total catch of gill nets with different mesh sizes that could be harvested if each hectare of water surface were caught once a year. Such fishing cannot adversely affect the stock status as the absolute catchability coefficient of nets is rather small and probably does not exceed 0.1. Thus, the biomass values in the table are in fact the total allowable catch (TAC) that can be targeted by the fishery. Thus, the biomass values in the table are essentially the total allowable catch that can be targeted by the fishery. Consequently, bottom-net fishing can yield a total catch of about 38 tons and corresponding catches of each species. In fact, it may be necessary to consider only the nets with the mesh size that are actually used in the fisheries, as well as allowable fishing intensity and deployment, but the calculation scheme remains the same.